United States
Department
of Agriculture

Forest Service

**Rocky Mountain
Research Station**

General Technical
Report RMRS-GTR-42-
volume 1

January 2000

Wildland Fire in Ecosystems

Effects of Fire on Fauna

Abstract

Smith, Jane Kapler, ed. 2000. Wildland fire in ecosystems: effects of fire on fauna. Gen. Tech. Rep. RMRS-GTR-42-vol. 1. Ogden, UT: U.S. Department of Agriculture, Forest Service, Rocky Mountain Research Station. 83 p.

Fires affect animals mainly through effects on their habitat. Fires often cause short-term increases in wildlife foods that contribute to increases in populations of some animals. These increases are moderated by the animals' ability to thrive in the altered, often simplified, structure of the postfire environment. The extent of fire effects on animal communities generally depends on the extent of change in habitat structure and species composition caused by fire. Stand-replacement fires usually cause greater changes in the faunal communities of forests than in those of grasslands. Within forests, stand-replacement fires usually alter the animal community more dramatically than understory fires. Animal species are adapted to survive the pattern of fire frequency, season, size, severity, and uniformity that characterized their habitat in presettlement times. When fire frequency increases or decreases substantially or fire severity changes from presettlement patterns, habitat for many animal species declines.

Keywords: fire effects, fire management, fire regime, habitat, succession, wildlife

The volumes in "The Rainbow Series" will be published during the year 2000. To order, check the box or boxes below, fill in the address form, and send to the mailing address listed below. Or send your order and your address in mailing label form to one of the other listed media. Your order(s) will be filled over the months of 2000 as the volumes are published.

☐ RMRS-GTR-42-vol. 1. **Wildland fire in ecosystems: effects of fire on fauna.**

☐ RMRS-GTR-42-vol. 2. **Wildland fire in ecosystems: effects of fire on flora.**

☐ RMRS-GTR-42-vol. 3. **Wildland fire in ecosystems: effects of fire on cultural resources and archeology.**

☐ RMRS-GTR-42-vol. 4. **Wildland fire in ecosystems: effects of fire on soil and water.**

☐ RMRS-GTR-42-vol. 5. **Wildland fire in ecosystems: effects of fire on air.**

Send to: _____

Name

_____

Address

Fort Collins Service Center

Telephone	(970) 498-1392
FAX	(970) 498-1396
E-mail	rschneider/rmrs@fs.fed.us
Web site	http://www.fs.fed.us/rm
Mailing Address	Publications Distribution Rocky Mountain Research Station 240 W. Prospect Road Fort Collins, CO 80526-2098

Wildland Fire in Ecosystems

Effects of Fire on Fauna

Editor

Jane Kapler Smith, Rocky Mountain Research Station, U.S. Department of Agriculture, Forest Service, Missoula, MT 59807.

Authors

L. Jack Lyon, Research Biologist (Emeritus) and Project Leader for the Northern Rockies Forest Wildlife Habitat Research Work Unit, Intermountain (now Rocky Mountain) Research Station, U.S. Department of Agriculture, Forest Service, Missoula, MT 59807.

Mark H. Huff, Ecologist, Pacific Northwest Research Station, U.S. Department of Agriculture, Forest Service, Portland, OR 97208.

Robert G. Hooper, Research Wildlife Biologist, Southern Research Station, U.S. Department of Agriculture, Forest Service, Charleston, SC 29414.

Edmund S. Telfer, Scientist (Emeritus), Canadian Wildlife Service, Edmonton, Alberta, Canada T6B 2X3.

David Scott Schreiner, Silvicultural Forester (retired), Los Padres National Forest, U.S. Department of Agriculture, Forest Service, Goleta, CA 93117.

Jane Kapler Smith, Ecologist, Fire Effects Research Work Unit, Rocky Mountain Research Station, U.S. Department of Agriculture, Forest Service, Missoula, MT 59807.

Cover photo—Male black-backed woodpecker on fire-killed lodgepole pine. Photo by Milo Burcham.

Preface _____

In 1978, a national workshop on fire effects in Denver, Colorado, provided the impetus for the "Effects of Wildland Fire on Ecosystems" series. Recognizing that knowledge of fire was needed for land management planning, state-of-the-knowledge reviews were produced that became known as the "Rainbow Series." The series consisted of six publications, each with a different colored cover, describing the effects of fire on soil, water, air, flora, fauna, and fuels.

The Rainbow Series proved popular in providing fire effects information for professionals, students, and others. Printed supplies eventually ran out, but knowledge of fire effects continued to grow. To meet the continuing demand for summaries of fire effects knowledge, the interagency National Wildfire Coordinating Group asked Forest Service research leaders to update and revise the series. To fulfill this request, a meeting for organizing the revision was held January 4-6, 1993, in Scottsdale, Arizona. The series name was then changed to "The Rainbow Series." The five-volume series covers air, soil and water, fauna, flora and fuels, and cultural resources.

The Rainbow Series emphasizes principles and processes rather than serving as a summary of all that is known. The five volumes, taken together, provide a wealth of information and examples to advance understanding of basic concepts regarding fire effects in the United States and Canada. As conceptual background, they provide technical support to fire and resource managers for carrying out interdisciplinary planning, which is essential to managing wildlands in an ecosystem context. Planners and managers will find the series helpful in many aspects of ecosystem-based management, but they will also need to seek out and synthesize more detailed information to resolve specific management questions.

— The Authors
January 2000

Acknowledgments _____

The Rainbow Series was completed under the sponsorship of the Joint Fire Sciences Program, a cooperative fire science effort of the U.S. Department of Agriculture, Forest Service and the U.S. Department of the Interior, Bureau of Indian Affairs, Bureau of Land Management, Fish and Wildlife Service, and National Park Service. We thank Marcia Patton-Mallory and Louise Kingsbury for persistence and support.

The authors are grateful for reviews of the manuscript from James K. Brown, Luc C. Duchesne, R. Todd Engstrom, Bill Leenhouts, Kevin C. Ryan, and Neil Sugihara; the reviews were insightful and helpful. Reviews of special topics were provided by David R. Breininger, John A. Crawford, Steve Corn, and Kevin R. Russell; their help strengthened many sections of the manuscript. We are thankful to Nancy McMurray for editing; Dennis Simmerman for assistance with graphics; Bob Altman for literature reviews of special topics; Loren Anderson, Steve Arno, Milo Burcham, Robert Carr, Chris Clampitt, Betty Cotrille, Kerry Foresman, Jeff Henry, Catherine Papp Herms, Robert Hooper, Dick Hutto, Bob Keane, Larry Landers, Melanie Miller, Jim Peaco, Dean Pearson, Rick McIntyre, Dale Wade, and Vita Wright for providing photographs or helping us locate them.

Contents

Summary

Fire regimes—that is, patterns of fire occurrence, size, uniformity, and severity—have been a major force shaping landscape patterns and influencing productivity throughout North America for thousands of years. Faunal communities have evolved in the context of particular fire regimes and show patterns of response to fire itself and to the changes in vegetation composition and structure that follow fire.

Animals' immediate responses to fire are influenced by fire season, intensity, severity, rate of spread, uniformity, and size. Responses may include injury, mortality, immigration, or emigration. Animals with limited mobility, such as young, are more vulnerable to injury and mortality than mature animals.

The habitat changes caused by fire influence faunal populations and communities much more profoundly than fire itself. Fires often cause a short-term increase in productivity, availability, or nutrient content of forage and browse. These changes can contribute to substantial increases in herbivore populations, but potential increases are moderated by animals' ability to thrive in the altered, often simplified, structure of the postfire environment. Fires generally favor raptors by reducing hiding cover and exposing prey. Small carnivores respond to fire effects on small mammal populations (either positive or negative). Large carnivores and omnivores are opportunistic species with large home ranges. Their populations change little in response to fire, but they tend to thrive in areas where their preferred prey is most plentiful—often in recent burns. In forests and woodlands, understory fires generally alter habitat structure less than mixed-severity and stand-replacement fires, and their effects on animal populations are correspondingly less dramatic. Stand-replacing fires reduce habitat quality for species that require dense cover and improve it for species that prefer open sites. Population explosions of wood-boring insects, an important food source for insect predators and insect-eating birds, can be associated with fire-killed trees. Woodpecker populations generally increase after mixed-severity and stand-replacement fire if snags are available for nesting. Secondary cavity nesters, both birds and mammals, take advantage of the nest sites prepared by primary excavators.

Many animal-fire studies depict a reorganization of animal communities in response to fire, with increases in some species accompanied by decreases in others. Like fire effects on populations, fire effects on communities are related to the amount of structural change in vegetation. For example, understory fires and stand-replacement fires in grasslands often disrupt bird community composition and abundance patterns for only 1 to 2 years, but stand-replacement fires in shrublands and forests cause longer lasting effects, which are initially positive for insect- and seed-eating species and negative for species that require dense, closed canopy. Bird abundance and diversity are likely to be greatest early in succession. When shrub or tree canopy closure occurs, species that prefer open sites and habitat edges decline and species that prefer mature structures increase.

Major changes to fire regimes alter landscape patterns, processes, and functional linkages. These changes can affect animal habitat and often produce major changes in the composition of faunal communities. In many Western ecosystems, landscape changes due to fire exclusion have changed fuel quantities and arrangement, increasing the likelihood of large or severe fires, or both. Where fire exclusion has changed species composition and fuel arrays over large areas, subsequent fires without prior fuel modification are unlikely to restore presettlement vegetation and habitat. In many desert and semidesert habitats where fire historically burned infrequently because of sparse fuels, invasion of weedy species has changed the vegetation so that burns occur much more frequently. Many animals in these ecosystems are poorly adapted to avoid fire or use resources in postfire communities.

In the past 10,000 years, fire in North American ecosystems has not operated in isolation from other disturbances, nor has it occurred independent of human influence. In many areas, however, fire has been prevented or excluded for nearly 100 years, a level of success that is not likely to continue. Collaboration among managers, researchers, and the public is needed to address tradeoffs in fire management, and fire management must be better integrated with overall land management objectives to address the potential interactions of fire with other disturbances such as grazing, flood, windthrow, and insect and fungus infestations.

L. Jack Lyon
James K. Brown
Mark H. Huff
Jane Kapler Smith

Chapter 1: Introduction

Effects of wildland fire on fauna show almost infinite variety. Previous authors have limited discussion of this subject to only a few vertebrate groups (Bendell 1974), specific biotic provinces (Fox 1983; Stanton 1975), or general summaries (Lyon and others 1978). This report surveys the principles and processes governing relationships between fire and fauna. We recognize that this approach has limitations. We focus almost entirely on vertebrates, particularly terrestrial mammals and birds, because the information on those groups is most complete and the principles best documented. (Fire effects on aquatic vertebrates are summarized in "Effects of Fire on Soil and Water," another volume in this Rainbow Series.) We describe fire effects on specific faunal populations and communities by way of example, not as a survey of all that is known. Those seeking a detailed description of fire effects on fauna are referred to books that discuss the subject in general, such as Whelan (1995, chapter 6) and Wright and Bailey (1982, chapter 4); reports about fire effects in specific geographic regions (for example, McMahon and deCalesta 1990; Viereck and Dyrness 1979); and summaries of fire effects on specific faunal groups (for example, Crowner and Barrett 1979; Lehman and Allendorf 1989; Russell 1999). The Fire Effects Information System, on the Internet at www.fs.fed.us/database/feis, provides detailed descriptions of fire effects on more than 100 North American animal species and nearly 1,000 plant species.

Fires affect fauna mainly in the ways they affect habitat. Repeated fires have been a major force shaping landscapes and determining productivity throughout North America for thousands of years, with the possible exception of some portions of West Coast rain forests. Climate, vegetation, Native Americans, and fire interacted in a relatively consistent manner within each biotic region of North America before the advent of disease and settlers from Europe (Kay 1998). Prior to modern agriculture, fire suppression, and urbanization, vegetation patterns in each region were shaped by fire regimes with characteristic severity, size, and return interval (Frost 1998; Gill 1998; Heinselman 1981; Kilgore 1981).

The animal species native to areas with a centuries-long history of fire can obviously persist in habitat shaped by fire; many species actually thrive because of fire's influence. How? Animals' immediate response to fire may include mortality or movement. It is influenced by fire intensity, severity, rate of spread, uniformity, and size. Long-term faunal response to fire is determined by habitat change, which influences feeding, movement, reproduction, and availability of shelter (fig. 1). Alteration of fire regimes alters landscape patterns and the trajectory of change on the landscape; these changes affect habitat and often produce major changes in faunal communities.

USDA Forest Service Gen. Tech. Rep. RMRS-GTR-42-vol. 1. 2000

1

Figure 1—Elk rest and graze in unburned meadow adjacent to area burned by crown fire, Yellowstone National Park. Fire and fauna have coexisted in this ecosystem for thousands of years. Photo by Rick McIntyre, courtesy of National Park Service.

This volume is organized on the premises that fire regimes strongly influence animal response to fire and that fire affects animals at every level of ecosystem organization. In this chapter, we describe the fundamental concepts of fire regimes and their effects on vegetation structure. Chapter 2 describes the role of fire in several North American vegetation communities prior to settlement by European Americans. Because the vegetation provides habitat for fauna, this chapter provides background for understanding examples used in later chapters. The next four chapters describe animal response to fire at four levels of organization: individual, population, community, and landscape. Chapter 7 surveys fire effects on wildlife foods. Finally, chapter 8 discusses management implications of fire-fauna relationships, particularly in light of past fire exclusion, and identifies information gaps and research needs. Scientific names of all animals described in this report are listed in appendix A. Appendix B lists scientific names of plants. Appendix C contains a glossary of technical terms.

Historic Perspective

Fire has influenced composition, structure, and landscape patterns of animal habitat for millennia, so it is reasonable to assume that animals have coexisted and adapted to periodic perturbations from fire. Records show that lightning starts more than 6,000 fires each year in the United States; surely this force was just as powerful and ubiquitous in past millennia as it is now (Pyne 1982). Prior to the 1500s, millions of Native Americans lived in North America. They used fire regularly for many purposes (Kay 1998). Only recently, since the advent of fire exclusion policy and other activities that strongly influence fire regimes, has fire's influence on fauna been intensely questioned and investigated (Kilgore 1976).

During the era of European settlement of North America, fire came to be viewed in some geographic areas as a hopelessly destructive event that could not be stopped. Early legislation promoting fire control responded primarily to the loss of lives and settlements and vast mortality of harvestable trees

2

USDA Forest Service Gen. Tech. Rep. RMRS-GTR-42-vol. 1. 2000

that occurred in large fires around the end of the 18th century (for example, the Peshtigo Fire of 1871). However, proclaimed reductions of elk, deer, bison, and other wildlife populations associated with these fire events were also important considerations in establishing fire control legislation (Brown and Davis 1973).

Resource protection and associated fire control began with the establishment of forest reserves throughout North America at the end of the 1900s. The reserves were established mainly to protect the land from abuse by timber and grazing interests, but early reports from the reserves singled out fire as the greatest threat to America's grasslands, forests, and wildlife (Komarek 1962). The primary fire-related mission of land management agencies was reinforced in the 1930s: to stop fires wherever possible, and to prevent large fires from developing (Moore 1974). From that time until the 1960s, most managers and many of the public viewed fire as an unnatural event and an environmental disaster. The land area under vigorous fire protection grew every year, and the resources assigned to fire suppression grew accordingly (Brown and Davis 1973).

Even in the early days of fire exclusion policy, there were dissenting voices. In 1912, ecologist H. H. Chapman recognized that longleaf pine in the South was adapted to grow and mature in the presence of repeated fires (Chapman 1912). Subsequent studies by other researchers found that controlled burning improved quality of ungulate forage (Green 1931) and improved, restored, and maintained habitat for certain game species, especially the northern bobwhite (Stoddard 1931, 1935, 1936). The scientific community was beginning to view fire as a natural process and a tool for wildlife habitat management, but many public and private land managers strongly resisted the concept (Schiff 1962).

During World War II, fire suppression capability declined. The disastrous 1943 drought-related fires in the Southeastern United States prompted major shifts in government policies (Schiff 1962). By the 1950s, controlled burning to reduce fuels and enhance habitat for specific wildlife species had become commonplace, but all other fires were vigorously controlled. Meanwhile, scientists began to report striking changes in plant community composition and structure associated with fire exclusion. Important functions of fire were described for ponderosa pine in the Pacific Northwest (Weaver 1943), California chaparral and ponderosa pine (Biswell 1963), Arizona ponderosa pine (Cooper 1960), Florida Everglades (Loveless 1959), and interior Alaska (Lutz 1956).

With the publication of the Leopold Report (Leopold and others 1963) on ecological conditions of National Parks in the United States, managers and the public began to see the benefits of fires in wildlands. The Leopold Report established the concept that wildlife habitat is not a stable entity that persists unchanged in perpetuity, but rather a dynamic entity; suitable habitat for many wildlife species and communities must be renewed by fire. Policy began to shift away from the assumption that all wildland fires are destructive (Pyne 1982). In 1968, the fire policy of the USDI National Park Service changed drastically as managers began to adopt the recommendations of the Leopold Report. Policy officially recognized fire as a natural process to be managed for maintaining ecosystems and improving wildlife habitat. Thus began the current era of fire management in which fire is recognized as an integral part of ecosystems, including those aspects relating to fauna (Habeck and Mutch 1973).

Fire Regimes

Knowledge of the ecological role of fire in past centuries and descriptions of significant changes in the role of fire over time are essential for communication among professionals and citizens interested in resource management. Nearly every North American ecosystem has been drastically changed from conditions of past millennia. Regardless of how fire might be managed in the future in various ecosystems, information about its past role is important. As Morgan and others (1994) said, "Study of past ecosystem behavior can provide the framework for understanding the structure and behavior of contemporary ecosystems, and is the basis for predicting future conditions."

Fire varies in its frequency, season, size, and prominent, immediate effects, but general patterns occur over long periods. These patterns describe *fire regimes*. The practice of organizing biotic information around fire regimes originated in North America around 1980 (Heinselman 1978, 1981; Kilgore 1981; Sando 1978). Descriptions of fire regimes are general because of fire's tremendous variability over time and space (Whelan 1995). Nevertheless, the fire regime is a useful concept because it brings a degree of order to a complicated body of knowledge. The fire regimes that have influenced North American ecosystems in an evolutionary sense are those of pre-Columbian times (prior to 1500), before diseases introduced by European explorers began to decimate populations of Native Americans (see Kay 1995). While knowledge of pre-Columbian fire regimes would be useful for understanding ecosystem patterns and processes today, little information is available from that era. Detailed information available about past fire regimes is mostly based on biophysical evidence, written records, and oral reports that encompass the time from about 1500

USDA Forest Service Gen. Tech. Rep. RMRS-GTR-42-vol. 1. 2000

3

to the mid- to late-1800s. This was a time before extensive settlement by European Americans in most parts of North America, before extensive conversion of wildlands for agricultural and other purposes, and before fire suppression effectively reduced fire frequency in many areas. In this volume, we refer to the fire regimes of the past several centuries as "presettlement" fire regimes.

Fire frequency and severity form the basis for the commonly referenced fire regime classifications described by Heinselman (1978) and Kilgore (1981). Two concepts, fire return interval and fire rotation, describe the frequency with which fires occur on a landscape. *Mean fire return interval* is the average number of years between fires at a given location. *Fire rotation*, called by some authors the *fire cycle*, is the number of years that would be required to completely burn over a given area.

Fire severity describes the immediate effects of fire, which result from the rate of heat release in the fire's flaming front and the total heat released during burning. Fire severity determines in large part the mortality of dominant vegetation and changes in the aboveground structure of the plant community, so Kilgore (1981) refers to severe fires in forests as "stand-replacement" fires. The concept of stand replacement by fire applies to nonforest as well as forest areas. Fires in vegetation types such as prairie, tundra, and savannah are essentially all stand-replacing because the aboveground parts of dominant vegetation are killed (and often consumed) by fire. Most shrubland ecosystems also have stand-replacement fire regimes because fire usually kills the aboveground parts of shrubs. In this report, we refer to the following four kinds of fire regime:

1. *Understory fire regime* (applies to forest and woodland vegetation types)—Fires are generally not lethal to the dominant vegetation and do not substantially change the structure of the dominant vegetation. Approximately 80 percent or more of the aboveground dominant vegetation survives fires.
2. *Stand-replacement regime* (applies to forests, shrublands, and grasslands)—Fires kill or top-kill aboveground parts of the dominant vegetation, changing the aboveground structure substantially. Approximately 80 percent or more of the aboveground dominant vegetation is either consumed or killed as a result of fires.
3. *Mixed-severity regime* (applies to forests and woodlands)—Severity of fire either causes selective mortality in dominant vegetation, depending on different species' susceptibility to fire, or varies between understory and stand-replacement.

4. *Nonfire regime*—Little or no occurrence of natural fire (not discussed further in this volume).

See "Effects of Fire on Flora" (also in the Rainbow Series) for further discussion of fire regimes and comparison of this fire regime classification with others.

The literature demonstrates great local variation in fire effects on habitat, even within small geographic areas with a single fire regime. Fires theoretically should spread in an elliptical pattern (Anderson 1983; Van Wagner 1969), but the shape of burned areas and the fire severity patterns within them are influenced by fluctuations in weather during fires, diurnal changes in burning conditions, and variation in topography, fuels, and stand structure. Variable and broken topography and sparse fuels are likely to produce patchy burns, while landscapes with little relief and homogeneous fuels may burn more uniformly. It is no wonder then that fires shape a complex mosaic of size classes, vegetation structure, and plant species occurrence across the landscape, and this variety has a profound influence on the animals that live there.

Changes in Vegetation Structure

For animals, the vegetation structure spatially arranges the resources needed to live and reproduce, including food, shelter and hiding cover. Some fires alter the vegetation structure in relatively subtle ways, for example, reducing litter and dead herbs in variable-sized patches. Other fires change nearly every aspect of vegetation structure: woody plants may be stripped of foliage and killed; litter and duff may be consumed, exposing mineral soil; underground structures, such as roots and rhizomes, may be killed (for example, in most coniferous trees) or rejuvenated (for example, in many grass and shrub species, aspen, and oak). In this section, we summarize postfire structural changes according to the fire regimes described above.

Understory Fire Regimes

Understory fires change the canopy in two ways: by killing or top-killing a few of the most fire-susceptible trees, and by killing or top-killing a cohort of tree regeneration, also selectively according to fire resistance. Understory fires also reduce understory plant biomass, sometimes in a patchy pattern. Although the structural changes caused by any one understory fire are not dramatic, repeated understory fires shape and maintain a unique forest structure identified by O'Hara and others (1996) as "old forest, single stratum." It is characterized by large, old trees, parklike conditions, and few understory trees (fig. 2).

4

Figure 2—Mature longleaf pine forest, typical of forest structure maintained by frequent understory fire, in Francis Marion National Forest, South Carolina. This kind of habitat favors many fauna species, included red-cockaded woodpecker, Bachman's sparrow, northern bobwhite, fox squirrel, and flatwoods salamander. Photo by Robert G. Hooper.

Stand-Replacement Fire Regimes

Grasslands—In grasslands, the prefire structure of the vegetation reasserts itself quickly as a new stand of grass springs up from surviving root systems. Standing dead stems and litter are reduced. The proportion of forbs usually increases in the first or second postfire year. In about 3 years the grassland structure is usually reestablished (Bock and Bock 1990), and faunal populations are likely to resemble those of the preburn community. Repeated fires can convert some shrublands to grass, and fire exclusion converts some grasslands to shrubland and forest.

Shrublands—In shrub-dominated areas, including sagebrush, chaparral, and some oak woodlands, stand-replacing fires top-kill or kill aboveground vegetation. Canopy cover is severely reduced, but initial regrowth usually increases cover of grasses and forbs. Dead woody stems often remain standing and serve as perch sites for songbirds, raptors, and even lizards (fig. 3). Burning increases seed visibility and availability for small mammals but also increases their visibility to predators. Because cover for ungulates is reduced by fire, some species do not use the abundant postfire forage. Shrubs regenerate from underground parts and seed. The length of time required to reestablish the shrubland structure varies, from 2 years in saw palmetto scrub (Hilmon and Hughes 1965) to more than 50 years in big sagebrush (Wright 1986).

Forests and Woodlands—In tree-dominated areas, stand-replacing fires change habitat structure dramatically. When crown fire or severe surface fire kills most of the trees in a stand, surface vegetation is consumed over much of the area, and cover for animals that use the tree canopy is reduced. Crown fires eliminate most cover immediately; severe surface fires kill the tree foliage, which falls within a few months. Stand-replacing fires alter resources for herbivores and their predators. The habitat is not "destroyed," but transformed: The fire-killed trees become food for millions of insect larvae and provide perches for raptors. Trees infected by decay before the fire provide nest sites for woodpeckers and then for secondary cavity nesters (birds and mammals). As

USDA Forest Service Gen. Tech. Rep. RMRS-GTR-42-vol. 1. 2000

5

Figure 3—Sagebrush 3 years after stand-replacing fire, east-central Idaho. Fire kills sagebrush but leaves dead stems that birds and reptiles use as perches. The photo shows early successional dominance by dense bluebunch wheatgrass. Photo by Loren Anderson.

these snags fall, other fire-killed trees decay and provide habitat for cavity nesters. For 10 to 20 years after stand-replacing fire, biomass is concentrated on the forest floor, as grasses and forbs, shrubs and tree saplings reoccupy the site. These provide forage and dense cover for small mammals, nest sites for shrubland birds, and a concentrated food source for grazing and browsing ungulates. In 30 to 50 years after stand-replacing fire, saplings become trees and suppress the early successional shrub and herb layers. The forest again provides hiding and thermal cover for ungulates and nesting habitat for animals that use the forest interior. The remaining fire-killed snags decay and fall, reducing nest sites for cavity-nesting birds and mammals but providing large pieces of dead wood on the ground. This fallen wood serves as cover for small mammals, salamanders, and ground-nesting birds. The fungi and invertebrates living in dead wood provide food for birds and small mammals (for example, see McCoy and Kaiser 1990).

In some northern and western coniferous forests, the initial postfire stand is composed of broad-leafed, deciduous trees such as aspen or birch. Conifer dominance follows later in succession. Some bird and mammal species prefer the broad-leafed successional stage to earlier and later stages of succession. As succession continues, conifers dominate and broad-leafed trees decay. This process creates snags and

adds to dead wood on the ground, enhancing habitat for cavity nesters and small mammals. It also creates openings that are invaded by shrubs and saplings. Dense patches of shrubs and tree regeneration in long-unburned forests provide excellent cover for ungulates. Birds (for example, crossbills, nuthatches, brown creeper, and woodpeckers), tree squirrels, and American marten find food, cover, and nest sites within the structure of the old-growth coniferous forest.

In some Southeastern forests, the roles of pine and hardwood tree species are reversed. Many Southeastern forests regenerate to pine immediately after stand-replacing fire. In the absence of repeated understory fires, these pine stands are invaded and eventually dominated by broad-leafed deciduous species such as American beech, hickory, and southern magnolia (Engstrom and others 1984; Komarek 1968). As in the hardwood-conifer sere of the Western States, each structural stage supports a somewhat different assemblage of wildlife.

Mixed-Severity Fire Regimes

In mixed-severity fire regimes, fires either cause selective mortality of fire-susceptible species in the overstory or alternate between understory and stand-replacement, with overlapping burn boundaries. The

6

USDA Forest Service Gen. Tech. Rep. RMRS-GTR-42-vol. 1. 2000

net result is a fine grained pattern of stand ages and structures across the landscape. This pattern is accentuated in areas where variable topography and microclimate influence fire spread. Through feedback of fuel patterns into subsequent fire behavior, the variety in fuels and stand structures resulting from mixed-severity fire perpetuate the complex mosaic of ages and structures.

Snags and Dead Wood

It would be difficult to overestimate the importance of large trees, snags, and dead, down wood to North American birds and small mammals. According to Brown and Bright (1997), "The snag represents perhaps the most valuable category of tree-form diversity in the forest landscape." Fire and snags have a complex relationship. Fires convert live trees to snags, but fires also burn into the heartwood of old, decayed snags and cause them to fall. Fire may facilitate decay in surviving trees by providing an entry point for fungi, which increases the likelihood that the trees will be used by cavity excavators. Fire may harden the wood of trees killed during a burn, causing their outer wood to decay more slowly than that of trees that die from other causes. This "case-hardening" process reduces the immediate availability of fire-killed snags for nest excavation but slows their decay after they fall.

It is difficult to identify fire-injured trees that are likely to become snags, and it is also difficult to determine which snags may have the greatest "longevity," that is, may stand the longest time before falling. In ponderosa pine stands in Colorado, for example, the trees most likely to become long-lasting

snags are underburned trees with moderate crown scorch that remain alive for at least 2 years after fire, a group that cannot be determined until 2 or 3 years after fire (Harrington 1996). According to Smith (1999), longevity of ponderosa pine snags is positively related to tree age and size at death. Fire-scarred trees may have greater longevity than trees never underburned (Harrington 1996).

The usefulness of snags to fauna is enhanced or reduced by the surrounding habitat, since cavity nesters vary in their needs for cover and food. Many cavity excavators require broken-topped snags because partial decay makes them easier to excavate than sound wood (Caton 1996). Some bird species nest only in large, old snags, which are likely to stand longer than small snags (Smith 1999). Pileated woodpeckers are an example. Some excavators and secondary cavity nesters prefer clumps of snags to individual snags, so the spatial arrangement of dead and decaying trees in an area influences their usefulness to wildlife (Saab and Dudley 1998).

Dead wood on the ground is an essential habitat component for many birds, small mammals (fig. 4), and even large mammals, including bears (Bull and Blumton 1999). Large dead logs harbor many invertebrates and are particularly productive of ants; they also provide shelter and cover for small mammals, amphibians, and reptiles. Fire both destroys and creates woody debris. While large, down logs are not always abundant in early postfire years, fire-killed trees eventually fall and become woody debris. Down wood from fire-killed trees often decays more slowly than wood of trees killed by other means (Graham and others 1994).

Figure 4—Vagrant shrew travelling in shelter of dead log, Lolo National Forest, western Montana. Large dead wood is an essential source of food and shelter for many small mammals. Photo by Kerry R. Foresman.

USDA Forest Service Gen. Tech. Rep. RMRS-GTR-42-vol. 1. 2000

7

Notes

_____

_____

_____

_____

_____

_____

Edmund S. Telfer

Chapter 2:
Regional Variation in Fire Regimes

To provide a context for discussion of fire effects on animals and their habitat, this chapter describes the vegetation, fire regimes, and postfire succession of several plant communities referred to in subsequent sections of this report. This description is not meant to be a complete survey of fire regimes in North America; such a survey is available in "Effects of Fire on Flora," also part of the Rainbow Series. Instead, it provides examples of plant communities and fire regimes throughout the continent, many of them described in earlier reviews, including Wright and Bailey (1982). These communities are divided according to the geographic regions used to describe fire effects on the flora in this series: northern ecosystems; eastern ecosystems, including the Great Plains; western forests; western woodlands, shrublands, and grasslands; and subtropical ecosystems.

Northern Ecosystems

Boreal Forest

Vegetation and Fire Regimes—The Boreal Forest was characterized in presettlement times by stand-replacement fire regimes, although understory fires were common in some dry forest types (Heinselman

1981; Johnson 1992). Most of the presettlement fire rotations reported in the literature were relatively short, ranging from 50 to 100 years (Heinselman 1981; Payette and others 1989; Wein 1993). Johnson (1992) determined that the fire rotation was between 40 and 60 years for Minnesota, Ontario, the Northwest Territories, and Alaska. Relatively short rotations probably occurred in dry continental interior regions; for example, a 39-year rotation in northern Alberta (Murphy 1985). Longer rotations occurred in floodplains (Heinselman 1981) and Eastern boreal forests (Viereck 1983). From 1980 to 1989, frequency of fires larger than about 500 acres (200 ha) in the Canadian boreal forest was greatest in the central part of the continent and decreased toward the east and northwest. Fires were more frequent during dry climatic periods than during wet periods (Clark 1988; Swain 1973).

Due to frequent fires in the Boreal Forest, there probably has been no time during the last 6,000 to 10,000 years when ancient or even old forest covered a high proportion of the region (Telfer 1993). Van Wagner (1978) and Johnson (1992) found that the distribution of forest area over age classes often approximates a negative exponential distribution, permitting prediction of the distribution of age class areas under various fire rotations. Based on this

USDA Forest Service Gen. Tech. Rep. RMRS-GTR-42-vol. 1. 2000

9

relationship, Johnson (1992) commented that a 40- to 60-year fire rotation, "...by definition, suggests that most (63 percent) stands will never live much beyond the age at which stand canopy closure occurs and very few will reach anything resembling old age."

Postfire Succession—Principal Boreal Forest trees include black spruce, white spruce, jack pine, and quaking aspen. All of these species regenerate well on burned sites. Most of the understory plants that occur in the Boreal Forest sprout from underground parts that can survive fire. Ahlgren (1974) does not consider any boreal shrub species likely to suffer substantial mortality due to burning.

Croskery and Lee (1981) examined plant regrowth at burned and unburned sites on a large May and June stand-replacing fire in northwestern Ontario. Existing trees, mostly black spruce and jack pine, were killed by the fire, and aboveground parts of shrubs and ground cover were mostly consumed. However, regrowth began immediately. By mid-July, ground cover in the burned area had rebounded to 50 percent of that in the unburned area, with an average of 14 species present compared to 21 on unburned sites. In the second growing season after fire, shrubs began to appear on the burn. By the fifth growing season, ground cover was 40 percent and mean height of deciduous species was 5 feet (1.5 m). Browse biomass was eliminated on severely burned areas for 2 years, then became available in small amounts. By the fifth year, browse was abundant.

Laurentian Forest

Vegetation and Fire Regimes—The Laurentian Forest constitutes a broad ecotone between the Eastern Deciduous Forest and Boreal Forest. It contains plant and animal species characteristic of both regions and some species, like eastern white pine, red pine, and red spruce, whose distributions are centered here. The forest consists of extensive pine forest and stands of northern hardwoods intermixed with eastern hemlock. Studies of charcoal and plant pollen in lake sediments show that fire has influenced species composition of the vegetation in the eastern portion of the Laurentian Forest during much of the past 10,000 years (Green 1986).

Overall, the most common kinds of fire in the Laurentian Forest were stand-replacement and mixed-severity fire, although understory fires occurred as well (Heinselman 1981). Stand-replacing fire predominated in jack pine, black spruce, and spruce-fir forests, with fire rotations in the 50- to 100-year range (Heinselman 1981). In red and white pine forests, mixed-severity fires predominated. Presettlement fire rotations in some coniferous forests were 150 to 300 years (Wein and Moore 1977). In Northern hardwood

forests, fire rotations may sometimes have exceeded 1,000 years. The proportion of early successional stand area was small at any given time (Telfer 1993). Many fires were large, estimated at 1,000 to 10,000 acres (400 to 4,000 ha) (based on Heinselman 1981).

Postfire Succession—With so many species of both boreal and southern affinities in the Laurentian Forest, many combinations of species form in postfire succession. Long fire rotations create extensive stands of mature and old hardwoods (American beech, birches, and maples). Stand-replacement fires are followed by a flush of shrubs and saplings, including red and sugar maple, paper and gray birch, alders, quaking aspen and bigtooth aspen, and cherry and shadbush species. White and red pines are also prominent, especially on sandy soils.

Early in succession, northern red oak and bur oak often intermix with less shade-tolerant hardwoods and pines. Pole-sized trees may be dense. Balsam fir and red spruce invade and gradually increase in dominance. On dry ridges, sugar maple, red maple, American beech, and oaks eventually dominate. On uplands, sugar maple, yellow birch, and American beech dominate the usually long-lasting mature stage. Eastern hemlock dominates on mesic sites with red spruce, yellow birch, paper birch, and occasional eastern white pine. One particularly volatile combination of species occurs in the northern Laurentian Forest and the southeastern fringe of the Boreal Forest. There balsam fir is a dominant species that supports outbreaks of spruce budworm; budworm-killed forest is highly flammable.

Eastern Ecosystems and the Great Plains

Eastern Deciduous Forest

Vegetation and Fire Regimes—The Eastern Deciduous Forest had understory and stand-replacing fire regimes in the centuries before settlement by European Americans. Lightning-caused fires were common in the mixed mesophytic hardwood forests of the Appalachian uplands and the Mississippi Valley (Komarek 1974). Because precipitation was plentiful in most years, the fires usually burned small areas. Some areas in this forest type burned frequently, including those near the bluegrass grasslands of Kentucky, which supported herds of bison (Komarek 1974). Historians and anthropologists now suggest that a substantial proportion of this deciduous forest was kept in early successional stages through shifting cultivation, firewood cutting, and extensive burning by agricultural tribes of Native Americans (Day 1953; MacCleery 1993). Annual burning in these areas created parklike stands of large, open-grown trees, a high proportion of which were fire-resistant oaks and eastern white pines. These hardwood forests had little understory and many openings.

Stand-replacement and mixed-severity fires shaped most of the pine forests of Eastern North America, particularly the extensive stands of eastern white and red pine along the northern periphery of the Midwestern States and in southern Ontario (Szeicz and MacDonald 1990; Vogl 1970) and the pitch pine and eastern redcedar forests on the Atlantic Coastal Plain (Wright and Bailey 1982).

Postfire Succession—Pines are common early successional species in the Eastern Deciduous Forest (Komarek 1974). Hardwood species with vigorous sprouting ability, especially oaks, also tend to dominate after fire. Increased prominence of oaks is one of the most common results of disturbance in this kind of forest (Williams 1989). Shade-intolerant species, including tuliptree and sweetgum, regenerate well on burned land (Little 1974). Many herbaceous species invade burned areas aggressively. In southern parts of the region, repeated burning leads to a mixed ground cover of grasses and legumes amid patches of trees (Komarek 1974). Without repeated disturbance, hardwood trees reoccupy the land with oaks in the vanguard. Continued absence of fire permits Eastern deciduous forests to be dominated by sugar maple, red maple, eastern hemlock, and American beech.

Southeastern Forests

Vegetation and Fire Regimes—The Southeastern Pine Region extends in a great arc from eastern Texas around the Appalachian uplands to Virginia. The vegetation is characterized by the "southern pines"—longleaf, slash, loblolly, shortleaf, and sand pines (Komarek 1974; Wright and Bailey 1982). These pine species tolerate and even depend upon fire to different degrees, while most hardwood species in the Southeast are suppressed by fire. Protection from fire enables hardwood forests to develop.

The Southeastern Pine Region has a high incidence of lightning strikes. Lightning and ignitions by aboriginal peoples caused understory fires in most longleaf pine forests every 1 to 15 years during presettlement times (Christensen 1988; Myers 1990). Since many of the grass and forb species associated with these forests also depend upon frequent fires (Frost and others 1986), cattlemen, farmers, and hunters continued burning the southern pine forests until the widespread adoption of fire suppression practices in the 1930s. By that time, intentional burning to improve wildlife habitat was already recognized as a management tool; by 1950 it was a common practice

(Riebold 1971). Longleaf pine dominated the Coastal Plain forests (except wetlands) until the early 1900s. Several factors, including alteration of the fire regime, have since favored dominance by loblolly and slash pines, which are somewhat less fire tolerant.

In eastern Oklahoma, shortleaf pine forests probably burned in large, low-severity understory fires at intervals of about 2 to 5 years prior to fire exclusion (Masters and others 1995).

The dominant vegetation in sand pine-scrub stands was killed or top-killed by fire every 15 to 100 years. One such fire burned 34,000 acres (14,000 ha) in 4 hours (Myers 1990). Maintenance of sand pine-scrub vegetation requires these infrequent, severe fires; more frequent fires can convert sand pine-scrub to longleaf pine (Christensen 1988).

Postfire Succession—The overriding impact of fire in the Southeastern Pine Region has been the maintenance of pine forest at the expense of hardwood forest. Relatively frequent understory fires shape a pine forest of variable density and well developed ground cover. Understory burning removes shrubs and small trees as sources of mast for wildlife, but it creates and maintains a vigorous understory of grasses, forbs, and fire-resistant shrubs (Wright and Bailey 1982). In the absence of fire, hardwood species invade the pine stands and deciduous forests develop. In much of the region, these are dominated by a mixture of oak and hickory in combination with many other deciduous species (Eyre 1980).

Prairie Grassland

Vegetation and Fire Regimes—The primeval prairie grasslands of North America stretched from the Gulf Coast in Texas north to central Alberta and from Illinois to western Montana. Precipitation increases from west to east, creating three north-south belts of vegetation—shortgrass prairie in the West, mixed prairie in the North and East, and tallgrass prairie in the Central and Eastern regions. Of all natural regions of North America, the Prairie Grassland has been most heavily impacted by human use. Tallgrass Prairie has been almost totally converted to agriculture. Development is somewhat less in westerly parts of the grassland. Substantial portions of the Shortgrass Prairie remain in use for cattle grazing.

The fire regime of the grasslands prior to settlement and development for agriculture was one of stand-replacing fires on a short return interval, every year in some areas (DeBano and others 1998; Wright and Bailey 1982). Ignitions due to lightning were common (Higgins 1984), and Native Americans ignited many fires (Wright and Bailey 1982). Prairie fires were often vast, burning into the forest margins and preventing tree invasion of grasslands (Reichman 1987).

USDA Forest Service Gen. Tech. Rep. RMRS-GTR-42-vol. 1. 2000

11

Postfire Succession—In prairie grasslands, burning maintains dominance by fire-adapted grasses and forbs. Fire also maintains the productivity of grasslands, supplying fresh, nutritious vegetation that is used by herbivores. Fire effects are strongly influenced by season of burning and moisture conditions. Burning outside the growing season causes little change in biomass yield or species dominance; fire during the growing season is likely to reduce yield and change species dominance. Postfire recovery is delayed if a site is burned during drought or, where annuals dominate, before seed set (DeBano and others 1998).

All grassland communities are subject to invasion by shrubs and trees in the absence of fire. Invading species include oaks, pines, junipers, mesquite, and aspen (Wright and Bailey 1982).

Western Forests

Rocky Mountain Forest

Vegetation and Fire Regimes—The Rocky Mountain Forest Region occupies inland mountain ranges and plateaus from New Mexico to Alberta and British Columbia. Vegetation patterns are complex and varied due to climatic differences that arise from variation in elevation and topography and the great latitudinal extent of the region. The forests are mainly coniferous. Important dominant trees include ponderosa pine, lodgepole pine, spruces, and firs. West of the Rocky Mountains, from Idaho north into British Columbia, Douglas-fir, western larch, and grand fir are dominant tree species.

Rocky Mountain forests in past centuries had a variety of fire regimes: understory, mixed-severity, and stand-replacement. At low elevations, understory fires maintained large areas of ponderosa pine and Douglas-fir in an open, parklike structure for thousands of years prior to the 1900s. Fires on these sites increased grass and forb production. Stand-replacing fires and complex mixed-severity fires were common in subalpine spruce-fir and lodgepole pine forests; understory fires also occurred, especially on dry sites.

Presettlement mean fire return intervals in Rocky Mountain forests ranged from less than 10 years (Arno 1976) to more than 300 years (Romme 1980). Forests with a multistoried structure, including dense thickets of young conifers, were more likely to experience stand-replacing fire than open, parklike stands. When ignition occurred in lodgepole pine forests, old stands were more likely to burn than young stands (Romme and Despain 1989).

Postfire Succession—Stand-replacing fires were unusual in ponderosa pine during presettlement times but are now more common because of increased

surface fuels and "ladder" fuels (shrubs and young trees that provide continuous fine material from the forest floor into the crowns of dominant trees). In presettlement times, repeated understory fire maintained an open forest with grasses and forbs on the forest floor and scattered patches of conifer regeneration. Fires occasionally killed large, old trees, creating openings where the exposed mineral soil provided a seedbed suitable for pine reproduction (Weaver 1974). Many forests were composed of multiple patches of even-aged trees.

Higher-elevation spruce-fir forests experience occasional stand-replacing fire. Conifer seedlings and deciduous shrubs sprout after being top-killed by fire and dominate regrowth within a few years after fire. Regenerating stands often produce large volumes of browse until the tree canopy closes, 25 or more years after fire. In the Northern Rocky Mountains, where lodgepole pine forests are mixed with spruce-fir, serotinous lodgepole pine cones open after being heated by fire. Fire thus simultaneously creates a good seedbed for pine and produces a rain of seed. The result is quick regeneration of lodgepole pine, often in dense stands.

Sierra Forest

Vegetation and Fire Regimes—Mixed conifer forests occur in the Sierra Nevada of California. Important species include Douglas-fir, incense cedar, sugar pine, white fir, and California red fir.

Sierra forests are famous for their groves of giant sequoia trees. Understory fires typically burned these forests at average intervals of 3 to 25 years. This fire regime produced an open structure with a grass and forb understory and scattered tree regeneration, similar to the structure of Rocky Mountain ponderosa pine forests.

Sierra Nevada forests also include ponderosa pine, with a presettlement regime of frequent understory fire; montane forests with a complex mixture of conifer species; and subalpine forests of lodgepole pine, whitebark pine, and California red fir. Montane and subalpine forests had a complex presettlement fire regime that included infrequent understory fire, mixed-severity fire, and stand-replacement fires of all sizes (Kilgore 1981; Taylor and Halpern 1991).

Postfire Succession—The understory fires characteristic of Sierra mixed conifer and ponderosa pine forests maintained open structures with little accumulation of debris on the ground (Kilgore 1981; Weaver 1974). Understory fire maintained dominance by pines and giant sequoias, with understory species including manzanita, deerbrush, wedgeleaf ceanothus, and bitter cherry (Wright and Bailey 1982). In the absence of

12

USDA Forest Service Gen. Tech. Rep. RMRS-GTR-42-vol. 1. 2000

fire, less fire-resistant species, including white fir and incense cedar, invade and develop into dense, tangled patches of young trees.

Pacific Coast Maritime Forest

Vegetation and Fire Regimes—The Pacific Coast Maritime Forest is the most productive forest type in the world (Agee 1993). The area is ecologically important because of the many species, including animals, that depend on old age classes of trees. The area is economically important because of its rapid rates of tree growth and biomass accumulation. The Pacific Coast region has wet winters and dry periods in the summer. In late summer, fire danger can become high, leading to stand-replacing crown fires with awesome intensity as described by Weaver (1974) for the 1933 Tillamook fire in Oregon.

Sitka spruce, western hemlock, and coast redwood dominate Pacific Coast Maritime forests. In past centuries, fires occurred infrequently in Sitka spruce and coastal forests of western hemlock, although most western hemlock forests show evidence that they were initiated following fire. Inland western hemlock forests probably burned in a regime of somewhat more frequent, mixed-severity fire. In redwood forests on relatively dry sites, fires of all kinds—understory, mixed-severity, and stand-replacement—were more common, occurring as frequently as every 50 years (Agee 1993).

Near the coast, long fire rotations in presettlement times resulted in a large proportion, probably about two-thirds, of the forest in mature and old age classes at any one time. There was thus ample habitat for flora and fauna that prefer or can survive in old growth.

Postfire Succession—Douglas-fir is important over much of the Pacific Coast Maritime Region because it is resistant to fire as an old tree, is able to colonize disturbed sites, and has a life span of several hundred years. On upland sites in the region, stand-replacing fire can be followed by dense shrub communities dominated by salmonberry, salal, red huckleberry, and vine maple (Agee 1993). Even where Douglas-fir becomes established immediately after fire, red alder may overtop it for many years (Wright and Bailey 1982). Postfire shrubfields sometimes persist indefinitely and sometimes are replaced by shade-tolerant conifers that regenerate beneath the shrub canopy.

Understory fires tend to eliminate most trees except large Douglas-fir and coastal redwood, if present (Agee 1993). Shade-tolerant trees regenerate under the remaining canopy. Mixed-severity fires produce gaps in which Douglas-fir regenerates and grows rapidly. Where redwood grows on alluvial sites, mixed-severity fire favors development of large, old redwood trees along with dense redwood regeneration.

Western Woodlands, Shrublands, and Grasslands _____

Pinyon-Juniper

Vegetation and Fire Regimes—Pinyon-juniper woodlands are dry, open forests occurring in the Southwestern United States. Prominent overstory species include the pinyon pines, Utah juniper, one-seed juniper, and alligator juniper. Pinyon-juniper woodland occupies elevations between higher oak woodlands and lower grass- and shrub-dominated areas (Wright and Bailey 1982). Because of the open nature of pinyon-juniper woodland, grasses and shrubs are prominent in the understory.

In presettlement times, stand-replacing fires probably occurred at intervals averaging less than 50 years in pinyon-juniper woodlands. Because of fire, areas with mature pinyon-juniper cover were somewhat restricted to locations with rocky soils and rough topography, which inhibited fire spread (Bradley and others 1992; Kilgore 1981; Wright and Bailey 1982). Where livestock grazing reduced herbaceous fuels, fire occurrence decreased and pinyon-juniper woodlands expanded. In mature, closed stands, fire spreads poorly because surface fuels are sparse. High winds and a high proportion of pine to juniper increase the potential for fire spread (Wright and Bailey 1982). Fire-caused tree mortality is likely to be great where fine fuels are dense or tumbleweeds have accumulated.

Postfire Succession—The impact of fire depends on tree density and the amount of grass and litter in the stand. For a few years after fire, pinyon-juniper woodlands present a stark landscape of dead trees and nearly bare soil. Annual plants become established in a few years. These are followed by perennial grasses and forbs. Invading plants often include weedy species, especially on bare soil. Junipers and shrubs typically reestablish in 4 to 6 years. After 40 to 60 years, the shrubs are replaced by a new stand of juniper (Barney and Frischnecht 1974; Koniak 1985). Development of a mixture of mature pinyons and junipers can require up to 300 years (West and Van Pelt 1987).

Chaparral and Western Oak Woodlands

Vegetation and Fire Regimes—Chaparral and western oak woodlands include broad-leafed shrub and tree species that are well adapted to fire. These plant communities occur in dry mountains and foothills throughout the Southwestern United States. The largest area is in southwestern California and the foothills of the Sierra Nevada, where chaparral is notorious for its frequent, fast-spreading, stand-replacing

USDA Forest Service Gen. Tech. Rep. RMRS-GTR-42-vol. 1. 2000

13

fires. Following fire, chaparral species sprout and also establish vigorously from seed. Many species have seed that germinates best after being heated by fire (DeBano and others 1998). In California chaparral, stand-replacing fires have occurred every 20 to 40 years for hundreds of years (Kilgore 1981). Fires were less frequent in Arizona chaparral, at higher elevations in California, and on northern aspects.

Oak woodlands are characterized by species that resprout vigorously after fire; Gambel oak dominates many such woodlands in Utah and Colorado. Oak woodlands had understory fire regimes with occasional stand-replacing fire in presettlement times (Wright and Bailey 1982). Fire frequency was reduced in areas where grazing reduced surface fuels (Bradley and others 1992).

Postfire Succession—Annual and perennial herbs flourish after fire in chaparral, along with seedling and resprouting shrubs. Herbs are gradually eliminated as the dense overstory of large shrubs matures (DeBano and others 1998).

Browse productivity in chaparral increases dramatically during the first 4 to 6 years after burning (Wright and Bailey 1982) but declines thereafter. For a decade or two after fire, chaparral is quite fire resistant (Wright 1986). Burning at 20- to 30-year intervals maintains a diverse mixture of species. If the fire return interval is longer, sprouting species will dominate, reducing plant species diversity.

In oak woodlands, fire either underburns or top-kills the dominant species and stimulates suckering at the bases of oaks. It thus changes the structure of oak woodlands, stimulates other shrubs, and produces a 2- to 3-year increase in productivity of grasses and forbs. Perpetuation of oaks and optimization of mast are wildlife management objectives in some locations because of widespread wildlife use of mast. In California, acorns are eaten by nearly 100 species of animals, including California quail, wild turkey, deer, and bear (McDonald and Huber 1995).

Sagebrush and Sagebrush Grasslands

Vegetation and Fire Regimes—Sagebrush dominates large areas in the Western United States in dense shrub stands and mixtures with grasslands. A common associate is antelope bitterbrush. Grasses and forbs are abundant. Sagebrush grasslands often intermix with forest cover, especially at higher elevations.

In presettlement times, fires burned sagebrush grasslands at intervals as short as 17 years and as long as 100 years (Wright and Bailey 1982). Fire severity in sagebrush varied, depending on the occurrence of sufficient grass and litter to carry fire. If fuel was sufficient, fires were stand-replacing and

severe, burning through the shrub crowns. Where fuels were sparse, fires were patchy. Varied patterns of vegetation and seasonal differences in burning conditions produced substantial differences in fire severity and effects.

Cheatgrass, an exotic annual, is favored by frequent burns, especially spring burns (Wright and Bailey 1982). Cheatgrass provides an accumulation of fine fuel that burns readily, so it alters the fire regime in sagebrush grasslands to much more frequent, stand-replacing fire (Kilgore 1981; Knick 1999). This disturbance reduces shrub cover severely and eliminates the patchy pattern formerly characteristic of sagebrush-dominated landscapes.

Postfire Succession—Fires in sagebrush grasslands reduce woody shrub species. Big sagebrush, a valuable wildlife browse species, is highly susceptible to injury from fire. Its recovery depends on season and severity of burn, summer precipitation, and frequency of burning. Big sagebrush may take more than 50 years to recover preburn dominance (Wright 1986). Antelope bitterbrush may be killed or only top-killed by fire, depending on the ecotype present and fire severity (Bedunah and others 1995).

Many grass and forb species thrive after fire and may delay regrowth of shrubs. Fires occurring every few years reduce perennial grasses and favor annuals, including cheatgrass. Shrubs reinvade during wet years. Sagebrush grasslands occasionally undergo severe droughts, which provide a major setback to shrub vegetation even in the absence of fire (Wright 1986).

Deserts

Vegetation and Fire Regimes—North American deserts occur in two separate areas of dry climate. The larger of the two areas extends from Baja California north through the Great Basin to central Idaho and Oregon (Humphrey 1974). This large region supports three floristically distinct deserts: the Sonoran Desert in the south, the Mojave Desert in southeastern California and southern Nevada, and the large Great Basin Desert to the north. The second North American desert area is the Chihuahuan Desert, located in the northern interior of Mexico and southern New Mexico.

In deserts with woody plants and tall cacti, fire severity in presettlement times depended on fuel loading and continuity. Severe fire was possible only after a moist, productive growing season; mixed-severity fire was more likely at other times. Fire was most frequent and widespread in the Great Basin Desert because of its greater shrub biomass (sagebrush) and because grass biomass was usually sufficient to carry fire between clumps of shrubs

14

USDA Forest Service Gen. Tech. Rep. RMRS-GTR-42-vol. 1. 2000

(Kilgore 1981). Next to the Great Basin Desert, the Chihuahuan Desert was the most prone to fire, while the Sonoran and Mojave Deserts only had enough ground cover to carry fire after occasional, unusually wet growing seasons (Humphrey 1974). In a review of fire effects on succulent plants, Thomas (1991) estimates that intervals between fires prior to European American settlement were as short as 3 years in some desert grasslands and more than 250 years in dry areas such as the Sonoran Desert.

Seasonal weather and grazing influence fire potential in deserts (Wright and Bailey 1982). A wet year produces large quantities of grasses and forbs, which provide fuel to carry fire. Grazing reduces these fine fuels, thus reducing potential fire spread.

Postfire Succession—Regrowth following fire depends on the availability of moisture. If burning is followed by a wet season, production of perennial grasses and some forbs may increase (Wright and Bailey 1982). In the most arid desert areas, fires may reduce density of shrubs and cacti for 50 to 100 years (Wright 1986). However, studies have shown substantial differences between species and also complex interactions among available moisture, grazing, and plant species (Cable 1967; MacPhee 1991; Wright and Bailey 1982). Several studies report increases in exotic annual grasses, including red brome and red stork's bill, after fire in desert ecosystems; both frequency and intensity of fires may have increased since the introduction of these grasses (Rogers and Steele 1980; Young and Evans 1973).

Subtropical Ecosystems _____

Florida Wetlands

Vegetation and Fire Regimes—The subtropical region of Florida is underlain by an expanse of limestone bedrock that is almost level and barely above sea level. Due to the area's flat surface and high annual rainfall—59 inches, 149 cm (Wright and Bailey 1982)—wetland covers much of the area. Lower places in the bedrock surface accumulate peat and support vegetation dominated by sawgrass. Where elevations are slightly higher, fresh water swamp or wet prairie vegetation occurs. Dry land occurs as knolls called "hammocks," which support mixed hardwood forest. Despite its extensive wetlands, fire has always influenced the ecology of southern Florida.

Postfire Succession—Burning has apparently maintained coastal marshes against mangrove invasion. Frequent fires in sawgrass kept fuel loadings low and prevented severe fires that would consume peat deposits. As peat accumulates, tree distribution expands out from the hammocks, increasing habitat for terrestrial fauna and decreasing habitat for wetland animals (Wright and Bailey 1982). Understory fires, occurring about five times per century on the average, maintained cypress stands by killing young hardwoods and suppressing hardwood regeneration (Ewel 1990). Severe fires after logging or draining swamps alter successional pathways, enhancing willows and eventual succession to hardwood forest.

USDA Forest Service Gen. Tech. Rep. RMRS-GTR-42-vol. 1. 2000

15

Notes

_____

_____

_____

_____

_____

_____

L. Jack Lyon
Edmund S. Telfer
David Scott Schreiner

Chapter 3:
Direct Effects of Fire and Animal Responses

This chapter summarizes current knowledge about the immediate and short term (days to weeks) effects of fire on terrestrial vertebrates: fire related mortality, emigration, and immigration. Within these topics, we describe fire effects mainly for two animal classes—birds and mammals. Information regarding reptiles, amphibians, and invertebrates is included if available in the literature.

Most animal species respond predictably to the passage of fire (Komarek 1969; Lyon and others 1978). These responses vary widely among species. Many vertebrate species flee or seek refuge, but some vertebrates and many insects are attracted to burning areas. Other behavioral responses to fire include rescuing young from burrows, approaching flames and smoke to forage, and entering recent burns to feed on charcoal and ash (Komarek 1969).

Injury and Mortality

Despite the perception by the general public that wildland fire is devastating to animals, fires generally kill and injure a relatively small proportion of animal populations. Ambient temperatures over 145 °F are lethal to small mammals (Howard and others 1959),

and it is reasonable to assume the threshold does not differ greatly for large mammals or birds. Most fires thus have the potential to injure or kill fauna, and large, intense fires are certainly dangerous to animals caught in their path (Bendell 1974; Singer and Schullery 1989). Animals with limited mobility living above ground appear to be most vulnerable to fire-caused injury and mortality, but occasionally even large mammals are killed by fire. The large fires of 1988 in the Greater Yellowstone Area killed about 1 percent of the area's elk population (Singer and Schullery 1989). Fire effects on habitat influenced the species' population much more dramatically than did direct mortality. Because of drought during the summer of 1988 and forage loss on burned winter range, elk mortality was high in the winter of 1988 to 1989, as high as 40 percent at one location (Singer and others 1989; Vales and Peek 1996).

Fire may threaten a population that is already small if the species is limited in range and mobility or has specialized reproductive habits (Smith and Fischer 1997). The now extinct heath hen was restricted to Martha's Vineyard for many years before its extirpation, where scrub fires probably accelerated its demise (Lloyd 1938).

USDA Forest Service Gen. Tech. Rep. RMRS-GTR-42-vol. 1. 2000

17

Season of burn is often an important variable in fauna mortality. Burning during nesting season appears to be most detrimental to bird and small mammal populations (Erwin and Stasiak 1979). Following the burning of a reestablished prairie in Nebraska, mortality of harvest mice in their aboveground nests was evident, and many nests of ground-nesting birds were found burned. Nestlings and juveniles of small mammals are not always killed by fire, however. Komarek (1969) observed adult cotton rats carrying young with eyes still closed out of an area while fire approached. While fire-caused mortality may sometimes be high for rodent species, their high reproductive potential enables them to increase rapidly in favorable environments and disperse readily into burned areas. Kaufman and others (1988b) describe this pattern for deer mouse and western harvest mouse populations in Kansas tallgrass prairie.

Birds

Fire-caused bird mortality depends on the season, uniformity, and severity of burning (Kruse and Piehl 1986; Lehman and Allendorf 1989; Robbins and Myers 1992). Mortality of adult songbirds is usually considered minor, but mortality of nestlings and fledglings does occur. In addition, a review by Finch and others (1997) points out that reproductive success may be reduced in the first postfire year because of food reductions from spring fires. Nest destruction and mortality of young have been reported for several ground-nesting species, including ruffed, spruce, and sharp-tailed grouse (Grange 1948), northern harrier (Kruse and Piehl 1986), and greater prairie-chicken (Svedarsky and others 1986). While eggs and young of ground-nesting birds are vulnerable to spring fires, long-term fire effects on bird populations depend partly on their tendency to renest. According to a review by Robbins and Myers (1992), wild turkeys rarely renest if their nests are destroyed after 2 to 3 weeks of incubation, while northern bobwhite may renest two or three times during a summer. For this reason, many biologists consider turkeys more vulnerable to fire. A mixed-grass prairie habitat in North Dakota was burned during the nesting season, but 69 percent of active clutches survived the fire and 37 percent eventually hatched. Nesting success was attributed in part to areas skipped by the fire as it burned in a mosaic pattern (Kruse and Piehl 1986). Underground nests, such as that of the burrowing owl, are probably safe from most fires.

In forested areas, fire effects on birds depend largely on fire severity. The young of birds nesting on the ground and in low vegetation are vulnerable even to understory fire during nesting season. Species nesting in the canopy could be injured by intense surface fire and crown fire, but this kind of fire behavior is more common in late summer than during the nesting season.

Mammals

The ability of mammals to survive fire depends on their mobility and on the uniformity, severity, size, and duration of the fire (Wright and Bailey 1982). Most small mammals seek refuge underground or in sheltered places within the burn, whereas large mammals must find a safe location in unburned patches or outside the burn. Lyon and others (1978) observe that small animals are somewhat more likely to panic in response to fire than large, highly mobile animals, which tend to move calmly about the periphery of a fire (fig. 5).

Most small mammals avoid fire by using underground tunnels, pathways under moist forest litter, stump and root holes, and spaces under rock, talus, and large dead wood (Ford and others 1999). Not all survive. Ver Steeg and others (1983), for instance, found numerous dead meadow voles after an early spring fire in Illinois grassland. Adequate ventilation inside burrows is essential for animal survival (Bendell 1974). Burrows with multiple entrances may be better ventilated than those with just one entrance (Geluso and others 1986). Small mammals living in burrows survived stand-replacing fire during summer in an ungrazed sagebrush-bunchgrass community in southeastern Washington (Hedlund and Rickard 1981). Most voles survived a prescribed burn in Nebraska grassland (Geluso and others 1986). Several retreated underground at the approach of the fire and returned to the surface after the fire had passed, apparently unharmed. Others remained aboveground, moving quickly through dense vegetation to outrun the fire. One individual sought refuge upon a raised mound of soil created by plains pocket gophers and was adequately sheltered there from heat and flame.

Small rodents that construct surface-level nests, such as brush rabbits, harvest mice, and woodrats (dusky-footed, desert, and white-throated), are more vulnerable to fire-caused mortality than deeper-nesting species, especially because their nests are constructed of dry, flammable materials (Kaufman and others 1988b; Quinn 1979; Simons 1991). Woodrats are particularly susceptible to fire mortality because of their reluctance to leave their houses even when a fire is actively burning (Simons 1991).

Direct fire-caused mortality has been reported for large as well as small mammals, including coyote, white-tailed deer, mule deer, elk, bison, black bear, and moose (French and French 1996; Gasaway and DuBois 1985; Hines 1973; Kramp and others 1983; Oliver and others 1998). Large mammal mortality is most likely when fire fronts are wide and fast moving, fires are actively crowning, and thick ground smoke

18

USDA Forest Service Gen. Tech. Rep. RMRS-GTR-42-vol. 1. 2000

Figure 5—Bison foraging and resting near burning area, Yellowstone National Park. Photo by Jeff Henry, courtesy of National Park Service.

occurs. Singer and Schullery (1989) report that most of the large animals killed by the Yellowstone fires of 1988 died of smoke inhalation. Because mortality rates of large mammals are low, direct fire-caused mortality has little influence on populations of these species as a whole (French and French 1996). Animal mortality, of course, provides food for scavenger fauna (fig. 6). The largest group of fire-killed elk in Yellowstone National Park was monitored for several months after it burned. Grizzly bears, black bears, coyotes, bald

Figure 6—Fire-killed deer after stand-replacing fire, Yellowstone National Park. Note "whitewash" on the deer's flanks, evidence of use by scavenging birds. Photo courtesy of National Park Service.

USDA Forest Service Gen. Tech. Rep. RMRS-GTR-42-vol. 1. 2000

19

eagles, golden eagles, and common ravens fed on the carcasses (French and French 1996). According to Blanchard and Knight (1996), the increased availability of carcasses benefited grizzly bears because drought had reduced other food sources.

The stand-replacing and mixed-severity fires of 1988 in the Greater Yellowstone Area, which occurred mostly in lodgepole pine-dominated forest, provided opportunities to study animal behavior during burns. Most thoroughly studied were large mammals, including bison, elk, bear, moose, and deer. French and French (1996) observed no large mammals fleeing a fire, and most appeared "indifferent" even to crowning fires. Singer and Schullery (1989) concluded that large mammals were sufficiently mobile to simply move away from danger during the fires. Bison, elk, and other ungulates grazed and rested within sight of flames, often 100 m or less from burning trees.

Reptiles and Amphibians

According to a review by Russell and others (1999), there are few reports of fire-caused injury to herpetofauna, even though many of these animals, particularly amphibians, have limited mobility. In a review of literature in the Southeastern States, Means and Campbell (1981) reported only one species that may suffer substantial population losses from fire, the eastern glass lizard. Another review (Scott 1996) mentions the box turtle as being vulnerable to fire, but there are many reports of box turtles and other turtle species surviving fires by burrowing into the soil (Russell and others 1999). No dead amphibians or reptiles were found after understory burning in a longleaf pine forest in Florida (Means and Campbell 1981). The vulnerability of snakes to fire may increase while they are in ecdysis (the process of shedding skin); of 68 eastern diamondback rattlesnakes marked before a fire in Florida, the only two killed were in mid-ecdysis.

Many reptiles and amphibians live in mesic habitat. Woodland salamanders in the southern Appalachians, for instance, use riparian sites and sites with plentiful, moist leaf litter. These sites are likely to burn less often and less severely than upland sites. The resulting microsite variation within burns may account for observations that fire has little effect on populations of these species (Ford and others 1999). Wetlands may provide refuge from fire, and activities such as breeding by aquatic species may be carried out with little interruption by fire (Russell and others 1999).

Many desert and semidesert habitats burned infrequently in past centuries because of sparse fuels. In these areas, as in mesic sites, patchy fire spread may protect herpetofauna from fire-caused injury and mortality. A growing concern is the conversion of vegetation in desert and semidesert, which burned infrequently in past centuries, to vegetation that now burns every few years due to invasion of weedy species (see "Effects of Altered Fire Regimes" in chapter 4). Animals in these ecosystems may not be adapted to avoid fire.

Invertebrates

The vulnerability of insects and other invertebrates to fire depends on their location at the time of fire. While adult forms can burrow or fly to escape injury, species with immobile life stages that occur in surface litter or aboveground plant tissue are more vulnerable. However, aboveground microsites, such the unburned center of a grass clump, can provide protection (Robbins and Myers 1992). Seasonality of fire no doubt interacts with phenology for many invertebrates. Research is needed on fire effects at all stages of insect life cycles, even though larval stages may be more difficult to track than adult stages (Pickering 1997).

An August understory burn in South Carolina forest reduced the soil mesofauna as measured the day after fire, but annually burned plots had generally higher populations of soil mesofauna than did plots that had not been burned in 3 years or more (Metz and Farrier 1971).

Escape and Emigration

A second popular concept regarding animals' response to fire is that they leave the area permanently as soon as fire is detected. While non-burrowing mammals and most birds do leave their habitat while it is burning, many return within hours or days. Others emigrate because the food and cover they require are not available in the burn. The length of time before these species return depends on how much fire altered the habitat structure and food supply.

Birds

Many birds leave burning areas to avoid injury. Some return to take advantage of the altered habitat, but others abandon burned areas because the habitat does not provide the structure or foods that they require to survive and reproduce. While some raptors are attracted to fire (see "Immigration" below), others move out of an area immediately after fire. After the large Marble-Cone fire in California, the spotted owls in Miller Canyon abandoned their habitat (Elliott 1985). Spotted owls in south-central Washington continued to use areas burned by understory fire but avoided stand-replacement burns, probably because their prey had been reduced (Bevis and others 1997). Structural features make recent burns unsuitable

20

USDA Forest Service Gen. Tech. Rep. RMRS-GTR-42-vol. 1. 2000

habitat for some species. Although stand-replacing fire in a Douglas-fir forest in western Montana favored birds that feed on insects, at least one insect feeder, the Swainson's thrush, abandoned the burn immediately (Lyon and Marzluff 1985), probably due to its need for cover.

Several studies report declines in bird abundance or species diversity in the first year or two after stand-replacing fire, but few reports are available for the months immediately following fire. After a late October fire in 1980 in coastal chaparral, California, fewer birds of all species were seen in November. Three months later, the bird population remained 26 percent below average (McClure 1981). The number of bird populations absent or declining in postfire years 1 and 2 has been reported to exceed the number of populations remaining stable or increasing after fires in Saskatchewan grassland (Pylypec 1991), Kansas shrub-grassland (Zimmerman 1992), California coastal sage scrub (Stanton 1986), and Wyoming spruce-fir-lodgepole pine forest (Taylor and Barmore 1980). Many bird species return to burned habitat 2 to 3 years after fire (Lyon and Marzluff 1985; Wirtz 1979).

Mammals

Because large mammals, such as moose and deer, depend on vegetation for forage, bedding, cover, and thermal protection, they abandon burned areas if fire removes many of the habitat features they need. Thus stand-replacing fires and understory burns that are severe enough to top-kill shrubs and young trees seem more likely to trigger high rates of emigration than patchy or low-severity fires. Woodland caribou in southeastern Manitoba avoided boreal forest burned by stand-replacing fire in favor of bog communities, lakes, and other unburned areas. Caribou may continue to avoid burns for 50 years or more, until lichens become reestablished in the new forest (Schaefer and Pruitt 1991; Thomas and others 1995). If recent burns provide some, but not all, habitat requirements for a species, the animals may stay near the edges of a burn. Immediately following large, stand-replacing fires in chapparal, Ashcraft (1979) reported mule deer grazing no farther than 300 feet (90 m) from cover.

Many small mammal species also leave burned habitats. Based on intensive trapping results, Vacanti and Geluso (1985) found that most voles survived a prescribed burn in Nebraska grassland but left the burned area until a new litter layer had accumulated, about two growing seasons later. Possible reasons for emigration included decreased protection from predators, decreased food availability, and more interactions among individuals. In the year after prescribed understory burns in conifer woodland with a shrubby understory in California, the abundance of small mammals was almost three times greater on unburned than burned plots, even though species composition did not vary significantly between burned and unburned areas (Blankenship 1982). Densities of western harvest mouse decreased the first year after tallgrass prairie was burned because their aboveground nests were destroyed and they left the area. During the same period, deer mice increased, apparently attracted by sparse ground cover that made seeds easy to find. Western harvest mouse densities in the burn increased the following spring and summer, with the populations on unburned sites serving as sources of dispersing individuals (Kaufman and others 1988b). In a southwestern Idaho shadscale-winterfat community, fire reduced the abundance of small mammals in the first postfire year. A decline in American badger numbers on the burn accompanied the small mammal decline (Groves and Steenhof 1988).

The effects of fire on mammal species are related to the uniformity and pattern of fire on the landscape. Fire has been cited by many authors as detrimental to American marten food and habitat (see Koehler and Hornocker 1977). However, a mixed-severity fire in an area of lodgepole pine, spruce, and fir in northern Idaho left a mosaic of forest types that supported a diversity of cover and food types favorable for marten (Koehler and Hornocker 1977). During summer and fall, American marten feed on ground squirrels, fruits, and insects in areas burned by stand-replacing fire. They require dense forest during most winters, but they use open forest during mild winters. Thus while large, uniform burns do not favor American marten, a mosaic of vegetation shaped in part by recent fire may do so.

Immigration

Many animals are actually attracted to fire, smoke, and recently burned areas. Some of the most interesting research regarding immigration in response to fire is in the field of insect ecology. The beetles of the subgenus *Melanophila* ("dark loving"), for instance, use infrared radiation sensors to find burning trees, where they mate and lay eggs (Hart 1998). Most birds and mammals that immigrate in response to fire are attracted by food resources.

Birds

A few bird species are attracted to active burns, and many increase in the days and weeks that follow fire. Parker (1974) reports that black vulture, northern harrier, red-shouldered hawk, and American kestrel were attracted to an agricultural (corn stubble) fire in Kansas. In the Southwest, raptor and scavenger species that are attracted to fire or use recent burns for

USDA Forest Service Gen. Tech. Rep. RMRS-GTR-42-vol. 1. 2000

21

hunting include northern harrier, American kestrel, red-tailed hawk, red-shouldered hawk, Cooper's hawk, and turkey and black vultures (Dodd 1988). After the large, severe Marble-Cone fire in California, western screech-owls moved into the burned area (Elliott 1985). Many species of birds were attracted to a 440-acre (180-ha) burn on the Superior National Forest, Minnesota. About 10 weeks after the fire, the area was alive with bird activity. Species included sparrows, American robin, barn swallow, common grackle, American kestrel, northern flicker, common raven, hairy woodpecker, great blue heron, eastern bluebird, and black-backed woodpecker (Stensaas 1989).

Predators and scavengers are often attracted to burns because their food is more abundant or more exposed than on unburned sites. During small prescribed burns in Texas bunchgrass and mesquite-grass stands, white-tailed hawks were attracted to grasshoppers chased from cover by the fires. Turkey vultures and crested caracaras fed on small mammals that had died in the fire (Tewes 1984). Stand-replacing and mixed-severity fire in a Douglas-fir forest in western Montana favored birds feeding on insects (Lyon and Marzluff 1985). Immediately after the fire, intense activity by wood-boring insects, parasites of wood borers, and predaceous flies occurred, accompanied by "almost frenetic" feeding by warblers and woodpeckers. In another study of grassland fire, American kestrel and red-tailed hawk increased after burning (Crowner and Barrett 1979). During a grassland fire in Florida, both cattle egrets and American kestrels foraged close to the flames. Apparently the egrets were attracted to vertebrates and invertebrates, and the kestrels were preying exclusively on insects as they flew out of the fire, into the wind (Smallwood and others 1982).

Several studies show that woodpeckers are particularly attracted to burned areas. Black-backed woodpeckers are almost restricted to standing dead, burned forests in the Northern Rocky Mountains (Caton 1996; Hutto 1995; Lyon and Marzluff 1985) (fig. 7). Schardien and Jackson (1978) found pileated woodpeckers foraging extensively on logs in an area in Mississippi that had burned 2 weeks earlier; an abundant food supply of wood-boring beetles appeared to be the primary attraction. Woodpeckers were attracted to a stand-replacement burn in coastal sage scrub, probably to feed on insects in the fire-killed cover (Moriarty and others 1985).

When small mammals are attracted to abundant new growth in the months following fire, predators and scavengers are attracted too. Abundant prey attracted golden eagles and peregrine falcons to recently burned areas in New Mexico and southern California (Lehman and Allendorf 1989). Following stand-replacing fire in chaparral, common raptors and ravens were studied for an increase in numbers.

Only ravens increased, probably because of increased scavenging opportunities (Wirtz 1979). In Great Basin and Chihuahuan Desert shrubsteppe, patchy burns probably favor species that require perches and cover above the ground (Bock and Bock 1990).

Mammals

Most mammals travel at least occasionally to seek food and shelter, and some make lengthy migrations every year. Mammal species can readily move into burned areas. Some use burned areas exclusively, and some use them seasonally or as part of their home range.

Reports of mammals moving into burned areas immediately after fire are mainly anecdotal. Lloyd (1938) describes movement of large animals into burned areas to seek protection from insects. In California chaparral, mountain lions are attracted to the edges of recent burns where deer tend to congregate (Quinn

Figure 7—Male black-backed woodpecker at nest hole in fire-killed lodgepole pine. Photo by Richard L. Hutto.

22

USDA Forest Service Gen. Tech. Rep. RMRS-GTR-42-vol. 1. 2000

1990). Crowner and Barrett (1979) report red fox hunting in a recent burn in an Ohio field.

Many studies describe movement by large mammals to recently burned areas because of food quantity or quality. Courtney (1989) reports migration of pronghorn to a northern mixed prairie in Alberta 2 months after a July fire. The pronghorn fed on pricklypear cactus, which was succulent and singed, with many of the spines burned off. The following spring, pronghorn moved into the burn because vegetation there began growing approximately 3 weeks earlier than on unburned range. When the Delta caribou herd had its calves in Alaska in 1982, the caribou preferred a recently burned snow-free area to an unburned snow-free area and a snow-covered area (Davis and Valkenburg 1983). Seven months after a stand-replacing fire in boreal forest, northern Minnesota, yearling moose had moved into the burn, apparently attracted by increased forage and a low-density resident moose population (Peek 1972, 1974). Moose density increased from 0.5 per square mile a few months after the fire to more than 2 per square mile two growing seasons after the fire. Moose temporarily left the area during the winter, when the forage that had sprouted in response to fire was covered with snow (Peek 1972).

Large mammals may move into burned habitat simply because of familiarity with the area before fire. A behavioral study of Alaskan moose after stand-replacing and mixed-severity fire indicated that increased use of burned areas depended heavily on prefire travel patterns and awareness by the moose population of the area (Gasaway and others 1989).

Visibility of predators may be another reason for large ungulates to move into burned areas. Desert bighorn sheep abandoned areas from which fire was excluded (Etchberger 1990). Mazaika and others (1992) recommend prescribed burning in the Santa Catalina Mountains, Arizona, to clear large shrubs and maintain seasonal diet quality for bighorn sheep.

Most small mammals are able to migrate readily in response to increased food supplies, so many species repopulate burns quickly after fire. Removal of litter and standing dead vegetation, rather than increased growth of vegetation, seemed to attract deer mice to burned prairie within 5 weeks of a spring fire (Kaufman and others 1988a). Increased food availability apparently outweighed the increased danger of predation (Kaufman and others 1988b). After fire in Arizona chaparral, recolonization was "rapid" for the species that prefer grassy habitat, including voles, pocket mice, and harvest mice (Bock and Bock 1990).

Two landscape-related aspects of fire, size and homogeneity, influence colonization and populations of small mammals on recent burns. Research by Schwilk and Keeley (1998) showed a positive relationship between deer mouse abundance and distance from unburned edge, perhaps in response to food provided by postfire annual plants growing in the middle of burned areas. The fires, which burned in California chaparral and coastal sage scrub, left some "lightly burned" patches in canyon bottoms. These refugia may have enabled small mammals to colonize severely burned sites during the first 6 months after fire (Schwilk and Keeley 1998).

Reptiles and Amphibians

Little is known about amphibian and reptile emigration and immigration after fire. A study of low-severity prescribed fires in hardwood-pine stands of the South Carolina Piedmont found no evidence that herpetofauna emigrated in response to fire (Russell 1999). Western fence lizards in chaparral take refuge under surface objects at the time of fire; after the fire, they invade the burned site from unburned patches (Lillywhite and North 1974). Komarek (1969) reports seeing southern diamondback rattlesnakes sunning themselves in areas blackened by recent fire. Frequent lightning-season fires promote growth of the bunchgrasses that flatwoods salamanders seek out for laying their eggs. Fire exclusion reduces the grasses in favor of closed slash or pond pine forest (Carlile 1997).

USDA Forest Service Gen. Tech. Rep. RMRS-GTR-42-vol. 1. 2000

23

Notes

_____

_____

_____

_____

_____

_____

_____

L. Jack Lyon
Mark H. Huff
Edmund S. Telfer
David Scott Schreiner
Jane Kapler Smith

Chapter 4:
Fire Effects on Animal Populations

The literature describing animals' behavioral responses to fire, discussed in chapter 3, is limited. Furthermore, short-term responses do not provide insights about the vigor or sustainability of the species in an area. Studies of animal populations and communities are more helpful in providing such insights. Most research regarding fire effects on fauna focuses on the population level, reporting changes in abundance and reproductive success of particular species following fire. Population changes are the net result of the behavioral and short-term responses discussed in chapter 3, plus longer term responses (years to decades).

Numerous population studies report abundance and density of animals in relation to fire, but information on productivity and other demographic factors may be essential for understanding population responses. Research on the threatened Florida scrub-jay provides an example. The scrub-jay requires scrub oak associations (myrtle, Chapman, and sand live oak, ericaceous shrubs, and saw palmetto), often in areas with open pine cover (less than 15 percent), where pine densities are kept low by frequent understory fires. The best vegetation for the jays consists of a mosaic of different age classes of scrub, most of which have burned within the last 20 years. Optimum scrub height is about 4.5 feet (1.5 m), interspersed with shorter scrub (Breininger and others in press; Woolfenden 1973). Without fire, the oaks become too tall and the habitat too dense for the Florida scrub-jay because predators are not easily seen (Breininger and others 1995). Florida scrub-jay densities in areas with tall shrubs are sometimes greater than in areas with optimum-height shrubs. However, jay mortality in tall scrub exceeds reproductive success; the jay is unable to sustain a population in tall scrub, as it can in shorter scrub (Breininger and others in press).

Changes in Animal Populations ___

Birds

Bird populations respond to changes in food, cover, and nesting habitat caused by fire. The season of burning is important to birds in two ways: Fires during the nesting season may reduce populations more than fires in other seasons; and migratory populations may be affected only indirectly, or not at all, by burns that occur before their arrival in spring or after their departure in fall.

Most raptor populations are unaffected or respond favorably to burned habitat. Fires often favor raptors by

USDA Forest Service Gen. Tech. Rep. RMRS-GTR-42-vol. 1. 2000

25

reducing hiding cover and exposing prey populations. When prey species increase in response to postfire increases in forage, raptors are also favored. Dodd (1988) describes beneficial effects from fire on populations of burrowing owl in desert grassland, sharp-shinned and Cooper's hawk in chaparral, and northern goshawk and sharp-shinned hawk in ponderosa pine forest.

Fire effects on insect- and plant-eating bird populations depend on alterations in food and cover. The canyon towhee, which eats insects and seeds, increased after stand-replacing fire in chaparral, foraging for food in the recent burn (McClure 1981). Wirtz (1977) reports that swallows, swifts, and flycatchers were more abundant over burned than unburned chaparral during the first postfire year. California gnatcatchers in coastal sage scrub, however, require the structure and cover provided by mature scrub. They avoid burns for the first 4 to 5 years after fire (Beyers and Wirtz 1997). In the northern Rocky Mountains, Hutto (1995) found 15 bird species more abundant in communities recently burned by stand-replacing fire than in other habitat; most were bark-probing insect eaters. On a site burned 19 years previously by stand-replacing fire in Olympic National Park, hummingbirds were probably more abundant than anywhere else in the area because the burn provided abundant nectar-producing forbs and shrubs and also open space for courtship (Huff and others 1985). After mixed-severity and stand-replacement burns in central Idaho, lazuli buntings and chipping sparrows, both seed eaters, were the most abundant songbirds (Saab and Dudley 1998). Fire in marshes usually increases areas of open water and enhances forage for shorebirds and waterfowl (Vogl 1967; Ward 1968).

Bird nest site selection, territory establishment, and nesting success can be affected by season of fire. Spring burns may destroy active nests (Ward 1968). Duck nesting success in mixed-grass prairie in North Dakota was significantly lower in areas burned in spring than fall (Higgins 1986). Blue-winged teal, northern shoveler, and American wigeon showed the lowest nesting success on spring burns. The differences were short-lived, however. Duck nesting response to fall- and spring-burned areas was similar in the third postfire year.

Nesting success also depends on the quality of the habitat before fire. Most birds nesting in areas burned by stand-replacing fire in the northern Rocky Mountains used broken-topped snags that were present before the fire (Hutto 1995). Many species of woodpeckers show substantial population increases and disperse into areas burned by stand-replacing fire (Hejl and McFadzen 1998; Hutto 1995; Saab and Dudley 1998). After mixed-severity and stand-replacement burns in central Idaho, nest abundance for nine cavity-nesting species increased through postfire year 4. On burned, unlogged sites, all species had nesting success above 50 percent, and three Forest Service-sensitive species had 100 percent success (table 1) (Saab and Dudley 1998).

Ground-dwelling bird populations are likely to be affected by fires of any severity, whereas canopy-dwelling populations may not be affected by understory fire.

Table 1—Success of cavity-nesting species after stand-replacing and mixed-severity fires in ponderosa pine/Douglas-fir forest in central Idaho (Saab and Dudley 1998).

Species	No. nests/km surveyed in 1996, all treatments*	Nesting success, unlogged stands	Characteristics of preferred nesting habitat
		Percent	
Lewis' woodpecker**	0.70	100	Highest nesting success on standard logged sites, selected the largest, most heavily decayed snags
Hairy woodpecker	0.58	92	Highest nesting success on unlogged sites
Northern flicker	0.40	75	Highest nesting success on wildlife logged sites, selected the largest, most heavily decayed snags
Western bluebird	0.63	60	Highest nesting success on wildlife logged sites
Mountain bluebird	0.64	56	Highest nesting success on unlogged sites
American kestrel	0.29	not reported	Nested mainly on standard logged sites, selected heavily decayed snags
European starling	0.13	100	
White-headed woodpecker**	0.03	100	Selected heavily decayed snags
Black-backed woodpecker**	0.10	100	Favor unlogged sites, locations with high tree density, selected hard snags

* 1996 was postfire year 2 for sites in mixed-severity burn, postfire year 4 for sites in stand replacing burn. Three treatments were studied: standard salvage logged; wildlife logged (approximately 50 percent salvaged logged); and unlogged.
** Species listed as sensitive by Forest Service in Regions 1, 2, 4, or 6.

After a fall fire on prairie in Saskatchewan, populations of ground-dwelling birds declined significantly. Savannah sparrows and clay-colored sparrows, the two most common species, were both adversely affected by the burn. These species rely on shrubs, specifically western snowberry and silverberry, for nesting habitat (Pylypec 1991). The year after fire, the abundance of breeding pairs in the burn was less than half the abundance in unburned areas. The third postfire year, savannah sparrows had recovered to a breeding pair abundance 68 percent of that on unburned sites, but clay-colored sparrow abundance had not changed substantially.

Woodpeckers generally nest in snags or in the forest canopy. Reports indicate that populations of woodpecker using forests with understory fire regimes tend to be unaffected by underburns. Thinning from below, designed to emulate understory fire in reducing fuels in an old-growth forest in Oregon, did not alter use of the site by pileated woodpeckers or Vaux's swifts, another bird that uses the tree canopy in old-growth forests (Bull and others 1995). Pileated woodpeckers' ability to use underburned sites probably depends on fire severity. Fires that reduce logs, stumps, and snags could have adverse effects by decreasing insect availability. The endangered red-cockaded woodpecker inhabits open longleaf, loblolly and shortleaf pine forests with few hardwoods in the midstory. Winter and growing season understory fires every 2 to 5 years are essential for retarding the development of a hardwood midstory in red-cockaded woodpecker habitat (Carlile 1997; U.S. Department of the Interior, Fish and Wildlife Service 1985) (fig. 8). If a hardwood midstory does develop, the woodpecker abandons its territory (Loeb and others 1992). The most abundant red-cockaded woodpecker populations now occur in areas with a long history of aggressive prescribed burning (Costa and Escano 1989).

Bird populations may exhibit some plasticity in relation to postfire habitat use and nest site selection. Brewer's sparrows and sage sparrows have been described as specifically requiring large patches of dense sagebrush (Knick and Rotenberry 1995; Wiens and Rotenberry 1981), but evidence from burned areas suggests some adaptability. The Brewer's sparrow population declined after fire in big sagebrush in Idaho; however, this decline was neither severe nor long-lived (Petersen and Best 1987). Return rates of banded male Brewer's and male and female sage sparrows the first 4 years after fire did not differ between burned and unburned areas, except the second year after fire when fewer male sage sparrows returned to the burn. The burn may have benefited the sage sparrow population indirectly, since new males used the burn to establish their territories. Nest placement by Brewer's sparrow was examined in big

sagebrush rangeland before and after a prescribed fire in southeastern Idaho (Winter and Best 1985). Before the burn, all nests were located in sagebrush canopies. The prescribed fire burned about 65 percent of the vegetation, leaving a mosaic of burned and unburned sagebrush. After fire, there was a significant shift in nest placement: 21 percent were placed close to the ground. Fire may have reduced the number of tall shrubs, influencing some sparrows to nest beneath shrubs to obtain cover and concealment. Eastern kingbird populations in Michigan forests show an adaptable response to stand-replacing fire. In undisturbed riparian areas, eastern kingbirds nest in woody vegetation, which provides foliage for concealment, but they also nest successfully in the charred trunks and branches of burned jack pines (Hamas 1983). Several nests occurred in cupped depressions formed by embers that burned into heartwood.

Figure 8—Prescribed fire to improve red-cockaded woodpecker habitat. Fire is backing past a cavity tree on the Osceola National Forest, Florida. Photo by Dale Wade.

USDA Forest Service Gen. Tech. Rep. RMRS-GTR-42-vol. 1. 2000

27

Fires influence bird populations indirectly by altering the populations of associated invertebrate species. Chigger infestation in the bird community increased in a chaparral stand during the months following stand-replacing fire. Feather mites were reduced, perhaps because silicon dioxide in the ash killed the mites (McClure 1981).

Mammals

Most of the literature describing fire effects on small mammal populations is from studies of stand-replacement and mixed-severity fire. Like birds, mammals respond directly to fire-caused changes in cover and food. Spring fires may impact mammal populations more than fires in other seasons because of limited mobility of young. The species with the most vulnerable young are small mammals, most of which also have high reproductive rates; if postfire habitat provides food and shelter for them, their populations recover rapidly.

Ream (1981) summarized information in 237 references about small mammals and fire. She concluded that populations of ground squirrels, pocket gophers, and deer mice generally increase after stand-replacing fire. Kaufman and others (1982) also report that the deer mouse population increased after fire. They found more deer mice on 1- and 2-year-old burns in tallgrass prairie than in unburned areas. In the same study, western harvest mice were more abundant on unburned sites. One year after stand-replacing fire in shrub-steppe habitat in Idaho, the total number of small mammals was lower in burned plots than in unburned plots (Groves and Steenhof 1988), and most of the animals in the burn were deer mice.

Rabbits, showshoe hare, red squirrel, northern flying squirrel, and voles generally avoid recent stand-replacement burns, according to Ream (1981). Shrews avoid burned areas from which most of the litter and duff have been removed. Of 25 animal populations common in chaparral brushlands, two were more abundant in mature, closed chaparral than in recently burned sites: Townsend's chipmunk and dusky-footed woodrat (Biswell 1989). Northern red-backed voles avoided a stand-replacement burn in black spruce for 1 year and finally established a resident population in postfire year 4, coinciding with the first year of berry production in the burn (West 1982). In the first year after stand-replacing fire in California grassland and chaparral, populations of agile kangaroo rat, California pocket mouse, deer mouse, and California mouse were either unchanged or greater on burned than unburned areas. Populations of brush mouse, western harvest mouse, and woodrat species decreased or disappeared in burned chaparral and grasslands (Wirtz 1977). Mixed-severity fire had little impact on populations of small mammals in pitch pine forests of the southern Appalachians (Ford and others 1999).

Animals that are dormant or estivating in underground burrows during and immediately after fire are particularly well protected from direct fire effects. Populations of Townsend's ground squirrels, dormant below ground at the time of stand-replacing fire in a sagebrush-grass community in southeastern Washington, seemed unaffected by the fire (Hedlund and Rickard 1981). Research after a stand-replacing fire in chaparral found that the only burrowing rodents, Heerman and agile kangaroo rats, were also the only rodents to survive in substantial numbers, probably because their burrows protected them from heat (Quinn 1979).

Population responses of small mammals to fire are related to fire uniformity. Most reports of woodrat responses to fire indicate that they usually suffer relatively high mortality because their nests are above ground (Simons 1991). However, populations of woodrats were "unexpectedly high" in burned areas observed by Schwilk and Keeley (1998). These burns left patches of "lightly burned" vegetation in California chaparral and coastal sage scrub, which may have provided refugia for woodrat populations.

Ungulate species often benefit from increased food and nutrition on recent burns. Because ungulates are sensitive to alterations in vegetation structure, however, their net response to fire depends on its severity and uniformity. In Lava Beds National Monument, northern California, mule deer populations were little affected by fire; home ranges were neither abandoned nor extended as a result of burning (Purcell and others 1984). Mule deer populations in chaparral burned by stand-replacing fire often increase, benefiting from increased availability of browse. Mule deer density in climax chaparral was estimated at 25 per square mile, while density in a severely burned area was 56 per square mile (Ashcraft 1979). Fawn production the second spring after burning was 1.15 fawns per doe compared to 0.7 fawns per doe in climax chaparral. Biswell (1961) reported an even more dramatic increase: deer density in chamise chaparral rose from 30 deer per square mile in unburned brush to 120 deer per square mile the first year after stand-replacing fire. Density decreased each year after that until it reached preburn levels in 5 to 12 years. In contrast, Stager and Klebenow (1987) report that mule deer preferred pinyon-juniper stands 24 and 115 years after stand-replacing fire to recently burned stands. The difference may be attributable to the drier conditions in pinyon-juniper, which slow vegetation recovery from fire.

Most other large ungulates either respond neutrally or positively to postfire changes in habitat. Elk rely on browse in seral shrub fields during winter and use dense, pole-sized forest heavily in fall (Irwin and Peek 1983). Moose also rely on seral shrubs in many areas, especially where shrubfields are interspersed with closed-canopy forest. In two areas converted from sagebrush dominance to grassland with shrub patches, pronghorn were present after fire but not before; they had been absent from one site for 60 years prior to the burn (Deming 1963; Yoakum 1980). Bison may avoid burned areas for a short time, until regrowth of forage begins (Moore 1972). Several studies indicate that bison prefer foraging in recently burned areas the summer after fire (Boyce and Merrill 1991; Shaw and Carter 1990; Vinton and others 1993) (fig. 9). White-tailed deer prefer to browse on recent burns if cover is close by. Management recommendations for white-tailed deer for specific geographic regions often list a maximum opening size or minimum distance to cover (for example, see Ivey and Causey 1984; Keay and Peek 1980).

Large carnivores and omnivores are opportunistic species with large home ranges. Their populations change little in response to fire, but they tend to thrive in areas where their preferred prey or forage is most plentiful—often, in recent burns. Fire has been recommended for improving black bear (Landers 1987) and grizzly bear habitat (Hamer 1995; Morgan and others 1994) (fig. 10). In Minnesota, enough early postfire plant communities must exist within a gray wolf pack's territory to support a surplus of deer, moose, and American beaver for prey (Heinselman 1973).

American beaver populations invade streamside habitat where fire has stimulated regrowth of aspen or willow species (Kelleyhouse 1979; Ream 1981). Burned areas in New York had more beaver colonies and a higher annual occupancy than unburned areas (Prachar and others 1988).

Fire may indirectly reduce disease rates in large mammal populations. Following a stand-replacing fire in spruce-lodgepole pine and bunchgrass mosaic in Glacier National Park, Montana, bighorn sheep tended to disperse, which may have reduced lungworm infections in the population (Peek and others 1985).

Reptiles and Amphibians

Fire-caused changes in plant species composition and habitat structure influence reptile and amphibian populations (Means and Campbell 1981; Russell and others 1999). In chaparral, reptiles were more abundant in recently burned areas than in areas with mature, dense cover. Individual populations responded to the developing structure of the vegetation (Simovich 1979). Species that preferred open sites increased slightly during the first 3 years after fire. During the same time, species that used or could tolerate dense

Figure 9—Bison grazing in area converted by stand-replacing fire from shrub-dominated to forb- and grass-dominated cover. Photo by Jim Peaco, courtesy of National Park Service.

vegetation decreased but were not eliminated. As the chaparral becomes a dense, mature layer, reptile abundance is likely to decrease.

Amphibians in forested areas are closely tied to debris quantities—the litter and woody material that accumulate slowly in the decades and centuries after stand-replacing fire. In forests of British Columbia, the proportion of nonmammalian vertebrates (mainly amphibians) using woody debris was positively correlated with the length of the fire rotation (Bunnell 1995).

Many herpetofauna populations show little response to understory and mixed-severity fire. After mixed-severity fire in pitch pine stands in the southern Appalachian Mountains, populations of woodland salamanders were generally unchanged (Ford and others 1999). Low-intensity underburns in hardwood-pine stands of the South Carolina Piedmont did not significantly alter species richness of herpetofauna;

USDA Forest Service Gen. Tech. Rep. RMRS-GTR-42-vol. 1. 2000

29

Figure 10—Grizzly bears foraging in lodgepole pine regeneration following stand-replacing fire, Yellowstone National Park. Photo by Jim Peaco, courtesy of National Park Service.

amphibians were significantly more abundant on burned plots due to greater numbers of Fowler's toad and red-spotted newts (Russell 1999). Although the slash pine habitat of the flatwoods salamander in Florida was underburned during winter, its breeding season, the population showed no sign of decline (Means and Campbell 1981).

In longleaf pine forests and slash pine plantations in the Florida sandhills, the threatened gopher tortoise (fig. 11) requires a sparse tree canopy and open, grassy ground cover for optimum food and nesting (Carlile 1997; Means and Campbell 1981), conditions that are provided by understory burning. Fires during the growing season may increase nest sites and enhance food supplies for new hatchlings (Carlile 1997). More than 300 other species use the gopher tortoise's burrow, including numerous arthropods, reptiles, and amphibians, so fire effects on the tortoise impact many other populations in the faunal community (Carlile 1997; Means and Campbell 1981; Russell and others 1999; Witz and Wilson 1991).

A review by Russell and others (1999) explains that fire in isolated wetlands usually increases areas of open water and enhances vegetation structure favored by many aquatic and semiaquatic herpetofauna.

Invertebrates

At least 40 species of arthropods are attracted to fires (Evans 1971), alerted by stimuli including heat, smoke, and increased levels of carbon dioxide. Many use burned trees for breeding. When the larvae hatch, they feed on the burned trees.

Soil protects most soil macrofauna and pupae of many insects from fire. The level of protection depends on depth of the organism and depth of heat penetration, which in turn depend on duff consumption (Schmid and others 1981). Insect abundance above ground decreases immediately after fire in prairies but then increases as fresh, young plant growth becomes available (Robbins and Myers 1992).

30

USDA Forest Service Gen. Tech. Rep. RMRS-GTR-42-vol. 1. 2000

Figure 11—Two gopher tortoises graze on new grass shoots after a prescribed fire. Photo by Larry Landers.

Effects of Altered Fire Regimes

Understory Fire Regimes

Exclusion of fire can cause changes in faunal abundance and community composition in forests adapted to understory fire, but studies designed to examine the long-term effects of fire exclusion are rare. In the Southeast, the red-cockaded woodpecker requires longleaf pine habitat with an open midstory, maintained in past centuries by frequent understory fire. When a dense hardwood midstory develops due to fire exclusion, the woodpecker abandons its territory (Loeb and others 1992). Bird populations were monitored for 15 years in a loblolly and shortleaf pine stand in northwest Florida, comparing a site underburned annually to one from which fire had been excluded (Engstrom and others 1984). After 15 years of fire exclusion, the unburned plot had 20 times more trees and less than one-third the ground cover of the annually burned plot. In the fire-excluded stand, the bird community changed continuously in response to structural changes. Species that require open habitat disappeared within 5 years of fire exclusion. During years 3 to 7, another group of species reached maximum numbers (common yellowthroat, indigo bunting,

eastern towhee, white-eyed vireo, and northern cardinal). After saplings began to mature in the understory, species associated with mesic woods were observed. Populations of canopy-dwelling birds such as the eastern wood-pewee, great crested flycatcher, blue jay, and summer tanager were affected little by 15 years of succession.

Saab and Dudley (1998) hypothesize the effects of three future fire regimes on ponderosa pine-Douglas-fir forests with presettlement fire return intervals of 5 to 22 years (table 2). High intensity, stand-replacing fires would favor seven of the 11 cavity nesting bird species studied and would negatively affect four species. Continued fire suppression, accompanied by increasing forest density, would favor only two species. The third possibility discussed is a combination of silvicultural treatment and prescribed fire, which theoretically would favor eight species and negatively affect only two species. The table offers a framework for testing whether management to replace a presettlement regime of frequent understory fire with a combination of thinning and management-ignited understory fire can produce benefits similar to those from presettlement fire regimes to the species listed. To assess potential changes throughout the faunal community, such a table would need to include at least all indicator species and species of special concern.

USDA Forest Service Gen. Tech. Rep. RMRS-GTR-42-vol. 1. 2000

31

Table 2—Predicted responses by cavity-nesting birds to three possible fire regimes compared with the presettlement low intensity, high frequency fire regime in Idaho ponderosa pine/ Douglas-fir forests (Saab and Dudley 1998), presented as a framework of hypotheses to be tested. + = more favorable than presettlement regime, 0 = no different, - = less favorable.

| | Potential new fire regime | | |
Bird species	High intensity stand-replacement fire	Complete fire suppression	Prescribed fire with stand management
American kestrel	+	-	+
Lewis' woodpecker	+	-	+
Red-naped sapsucker	-	0	+
Downy woodpecker	-	0	+
Hairy woodpecker	+	0	+
Black-backed woodpecker	+	-	0
White-headed woodpecker	-	-	+
Northern flicker	+	+	-
Pileated woodpecker	-	+	-
Western bluebird	+	-	+
Mountain bluebird	+	-	+

Stand-Replacement Fire Regimes

Fire exclusion from areas with stand-replacing fire regimes has contributed to loss of habitat and population declines in several raptor and predator species. Examples include the golden eagle in the Appalachian Mountains (Spofford 1971) and short-eared owl along the eastern grassland-forest interface (Lehman and Allendorf 1989). A review by Nichols and Menke (1984) explains that several raptors (red-tailed hawks, Cooper's hawks, sharp-shinned hawks, and great horned owls) are more abundant in recently burned chaparral than in unburned areas due to greater visibility of prey.

Frequent, stand-replacing fire in presettlement times maintained a "virtually treeless landscape" on the Great Plains (Bidwell 1994). Fire exclusion, tree planting, and enhancement of waterways have encouraged woodlands to develop, fragmenting the prairies. The greater prairie-chicken, Henslow's sparrow, and upland sandpiper all decline where habitat is fragmented (Bidwell 1994). To increase abundance of the greater prairie-chicken in North Dakota, Kobriger and others (1988) recommend use of prescribed burning. To maintain nest and brood habitat for the prairie-chicken, Kirsch (1974) recommends burning large plots (at least 0.5 mile, 800 m, across) at 3- to 5-year intervals. Grasslands left undisturbed for more than 10 years are not desirable.

Prairie dog colonies once covered hundreds of thousands of acres of the Great Plains that burned frequently (Bidwell 1994). Prairie dogs prefer burned to unburned areas for feeding and establishing colonies (Bone and Klukas 1990). The prairie dog is essential prey for the black-footed ferret, and prairie dog colonies provide for needs of more than 100 other vertebrate species in some way (Scott 1996; Sharps and Uresk 1990). Prairie dog grazing and waste alter the soil and vegetation near colonies, favoring early successional forb species, stimulating growth of grass and forbs, and increasing the nitrogen content of forage (Bidwell 1994; Sharps and Uresk 1990). Improved grass forage attracts bison, and increased forb cover attracts pronghorn. Bison, in turn, trample the areas where they graze (Yoakum and others 1996). Bison impact on prairie dog colonies is reduced when recent burns are available for grazing (Coppock and others 1983).

Invasion by nonnative annual plants has increased fire frequency in many semidesert ecosystems that were characterized by stand-replacement fire regimes in presettlement times. Exotic annuals, particularly cheatgrass in sagebrush ecosystems, increase fuel load and continuity. The result is increased fire frequency, followed by greater area of bare soil that is colonized by greater numbers of exotic annuals (U.S. Department of the Interior 1996; Whisenant 1990). The impact of exotic annuals is exacerbated in sagebrush ecosystems because fire exclusion and overgrazing since the mid-1800s increased sagebrush dominance at the expense of native herbaceous species. Loss of sagebrush cover and disruption of the historic balance of shrubs, native grasses, and forbs threatens the viability of sage grouse, sage sparrow, Brewer's sparrow, and sage thrasher populations (Knick and Rotenberry 1995; Sveum and others 1998). In the Snake River Birds of Prey National Conservation Area, big sagebrush has declined from more than 80 percent cover in the 1800s to less than 15 percent in 1996 (U.S. Department of the Interior 1996). Areas that have burned in the last 15 years

32

USDA Forest Service Gen. Tech. Rep. RMRS-GTR-42-vol. 1. 2000

have less than 3 percent sagebrush cover. Models predict complete loss of shrublands in 25 to 50 years without fire suppression in cheatgrass areas. Loss of sagebrush is contributing to a steady decline in black-tailed jackrabbit populations and increased fluctuations in Townsend's ground squirrel populations. Prairie falcons and golden eagles rely on these two prey species, so increased fire frequency is reducing the density and reproductive success of both species (Wicklow-Howard 1989; U.S. Department of the Interior 1996). Other animals in Idaho that prey on the Townsend's ground squirrel—red-tailed hawks, American badgers, western rattlesnakes, and gopher snakes—may also be affected (Yensen and others 1992).

Lodgepole pine and aspen communities in the Western States provide two examples of effects of fire exclusion on forests with stand-replacement fire regimes. In lodgepole pine-spruce-fir forests, the most productive period for bird communities appears to be the first 30 postfire years. Thirteen species regularly breed only in the first 30 years after fire. Conversely, just two species breed exclusively in forests more than 30 years old (Taylor 1969, 1979; Taylor and Barmore 1980). Species that breed exclusively in the first 30 years after fire may be difficult to maintain in the ecosystem without fire. Fire exclusion and postfire salvage of dead trees after fire may reduce populations of these species over large geographic areas.

Aspen stands provide more forage and a greater diversity of understory plants than the spruce and fir communities that generally replace them in the absence of fire. Fires of moderate to high severity can regenerate aspen, but the moderate to high intensity fire necessary to stimulate vigorous suckering of aspen is often difficult to achieve (Brown and DeByle 1982; Severson and Rinne 1990).

DeMaynadier and Hunter's (1995) review points out that most research on effects of fire on amphibians and reptiles has been done in Florida. They and other authors (Russell and others 1999) caution against extending the results of this research to ecosystems where frequent fire was not part of the presettlement disturbance regime.

Mixed-Severity Fire Regimes

Not enough information is available to generalize about effects of changing fire regimes in areas with presettlement patterns of mixed-severity fire. Exclusion of fire from mixed-conifer and Douglas-fir forests in the Southwest has led to increased fuel loads and increasing risk of large, uniformly severe fire (Fiedler and Cully 1995; Lehman and Allendorf 1989; U.S. Department of the Interior, Fish and Wildlife Service 1995). Severe fire is likely to destroy nesting trees

and the dense forest structure required by Mexican and California spotted owls. Prescribed understory fire has been recommended to reduce fuels in areas near spotted owl nest trees and to break up fuel continuity in large areas of continuous dense forest, reducing the likelihood of large, stand-replacing fires in the future (Fiedler and Cully 1995; U.S. Department of the Interior, Fish and Wildlife Service 1995; Weatherspoon and others 1992).

Grazing and fire exclusion have converted some desert grasslands to open woodlands. This constitutes loss of habitat for species such as pronghorn and Ord's kangaroo rat but increases habitat for mule deer (Longland 1995; MacPhee 1991).

Animal Influences on Postfire Habitat

Most of the literature regarding the relationship between fire and fauna focuses on fire-caused changes in vegetation and how habitat changes influence animal populations. A related topic is the effect of animal populations on the process of postfire succession. In this brief section, we provide a few examples of such relationships for animals and plants native to North America.

The jay-sized Clark's nutcracker (fig. 12) is responsible for most whitebark pine regeneration (Tomback 1986). The nutcracker prefers to cache seed in open sites with highly visible landmarks, conditions available within recent burns (Murray and others 1997) (fig. 13). Tomback and others (1996) studied whitebark pine regeneration after the 1988 fires in the Greater Yellowstone area. Areas burned by mixed-severity

Figure 12—Clark's nutcrackers cache seed of whitebark pines. Unrecovered seed from these caches accounts for most whitebark pine regeneration. Photo by Bob Keane.

USDA Forest Service Gen. Tech. Rep. RMRS-GTR-42-vol. 1. 2000

33

Figure 13—Whitebark pine regeneration in an area burned by stand-replacing fire 30 years previous to the photo. Photo by Stephen F. Arno.

and stand-replacement fire had greater whitebark pine regeneration than did unburned sites.

Bison not only prefer burned to unburned grassland for grazing during the growing season, they also contribute to the pattern of burning in prairie. In tallgrass prairie in northeastern Kansas, bison selected patches with low forb cover dominated by big bluestem, and grazed larger patches in burned than unburned habitat (Vinton and others 1993). Ungrazed forbs in areas adjacent to heavily grazed patches were thriving, producing greater biomass than in larger, ungrazed portions of the study area. The increased variability in vegetation productivity may act as feedback to fire behavior, increasing variation in patchiness and variable severity of subsequent fires. During the centuries before European American settlement, bison populations may have been controlled by Native American hunting, which would have reduced the effects of grazing on fuel continuity (Kay 1998).

Kangaroo rats and pocket mice may enhance postfire dominance of Indian ricegrass in sagebrush grassland ecosystems. These rodent species gather and hoard large numbers of seed, with a clear preference for Indian ricegrass. On burned sites with abundant populations of these rodents, Indian ricegrass seed had been deposited in scatter-hoards before fire even though the species was not dominant. Indian ricegrass dominated soon after fire. Six years after fire, density of Indian ricegrass was more than tenfold greater on burned than unburned sites (Longland 1994, 1995).

Although fire causes high mortality for antelope bitterbrush, it also creates litter-free sites, in which bitterbrush germination rates are high. Most antelope bitterbrush seedlings originate in rodent seed caches, and rodents apparently retrieve fewer seed from sites with limited cover (such as burned areas) than from sites with better protection (Bedunah and others 1995; Evans and others 1983).

Grazing and browsing on postfire sites, whether by wild ungulates or domestic grazers, can alter postfire succession. For example, if aspen is treated by fire to regenerate the stand but then repeatedly browsed by wildlife, it often deteriorates more rapidly than without treatment (Bartos 1998; Basile 1979). Such intense effects of feeding by large ungulates only occur where the animal populations are food limited. Where Native American predation kept populations of these animals in check, such effects are unlikely (Kay 1998).

Mark H. Huff
Jane Kapler Smith

Chapter 5:
Fire Effects on Animal Communities

Many animal-fire studies depict a "reorganization" of animal communities resulting from fire, with increases in some species accompanied by decreases in others. Descriptions of faunal communities after fire, however, are much less prevalent than descriptions of population changes. The literature about fire and bird communities is more complete than the literature about other kinds of animals. In this chapter, we use the literature about fire and birds to search for response patterns in the relationship between fire regime and changes in bird community composition. The literature does not at this time provide enough studies of mammal communities to complete a similar analysis.

Each animal species in a community is likely to respond differently to fire and subsequent habitat changes. To synthesize information about these responses, we modified Rowe's (1983) classification of plant responses to fit animal responses to fire. Rowe's approach was to assign to each plant species an adaptation category based on reproduction and regeneration attributes in the context of fire. Using similar categories in our evaluation of the animal-fire literature, we classified species' responses (*not* species themselves) for a given study location using observed changes in animal abundance. Mean changes in species abundance before and the first few years after fire, or in burned versus unburned areas, can be classified into one of six categories (table 3). Possible community response patterns using these six categories include:

A. **Increasers predominate**: A high proportion of invader and/or exploiter responses. This pattern represents an upward shift in abundance, especially for opportunistic species.

B. **Decreasers predominate**: A high proportion of avoider and/or endurer responses. This pattern represents a downward shift in abundance and unsuitable or poor habitat conditions for species established prior to the burn.

C. **Most populations change**: An equitably high proportion of invader and/or exploiter responses and of avoider and/or endurer responses. This pattern represents a small change in total abundance but a large shift in abundance of many individual species.

D. **Few populations change**: A high proportion of resister responses and a low proportion of other responses. This pattern represents little change in species composition and relatively minor fire effects on the animal community.

E. **Intermediate change**: A high proportion of resister, endurer, and exploiter responses; low proportion of invader and avoider responses.

Table 3—Classification of changes in bird abundance into response categories.

Response category	Before fire	After fire
Invader	Not detected	Detected (minimum number)
Exploiter	Detected	>50% increase
Resister	Detected	≤50% increase or decrease
Endurer	Detected	>50% decrease
Avoider	Detected	Not detected or very low numbers
Vacillator	Detected/not detected	Inconsistent, wide fluctuations

This chapter presents bird community responses to fire according to the fire regimes as described in chapter 1: understory, stand-replacement, and mixed-severity. Understory fire regimes occur only in forest cover types. Stand-replacing fire regimes are divided according to vegetation type: grassland, shrub-grassland, shrubland, and forest. Finally, we discuss mixed-severity fires (also limited to forest types) that leave at least 40 percent cover of large trees.

Analysis of the literature using the framework described above shows that fire effects on bird communities are related to the amount of structural change in vegetation. In burned grasslands, bird communities tend to return to prefire structure and composition by postfire year 3. Postfire shrub communities are generally in flux until the shrub canopy is reestablished, often 20 years or more after fire. In forests, understory fire usually disrupts the bird community for 1 year or less. Stand-replacing fire generally alters bird communities for 30 years or more. However, variation is great. Many bird communities conform only loosely to this pattern.

Many studies of fire effects on bird communities report species richness or other indices of diversity. Conserving all species is obviously essential for sustaining ecosystem patterns and processes, but *maximizing* diversity in a given location does not necessarily sustain the ecosystem (Telfer 1993). Bird responses to fire in Southeastern scrub communities provide an example. Many bird species (for instance, the Carolina wren and northern cardinal) are negatively affected by regimes of frequent fire in these scrub communities. Increasing fire frequency may reduce these species, thus reducing species richness. But the populations reduced by frequent fire represent forest edge species common in Eastern North America. In contrast, increasing fire frequency favors the threatened Florida scrub-jay and other scrub specialists, which have a narrow geographic range and are the species that make Florida scrub habitat unique (Breininger and others in press). Their habitat is declining because fire frequencies have declined, and these changes have long-lasting effects on habitat structure even when fires later return to the system (Duncan and others 1999).

Understory Fire Regimes

Understory fires burn beneath the tree canopy, mostly through surface and understory fuels. Prescribed understory burns are commonly used to reduce fuel hazards and maintain open forest structure in areas that had high-frequency, low-intensity fire regimes in presettlement times, such as southeastern pines and ponderosa pine (see Biswell 1989). Understory fires often disrupt the bird community during the first postfire year, but by postfire year 2, underburned forests are generally returning to preburn bird community structure and composition.

The time since burn and the interval between understory fires influence fire effects on bird populations. In oak scrub and slash pine communities along the central east coast of Florida, for example, Carolina wren and white-eyed vireo had highest densities in areas that had not burned for more than 10 years. Common yellowthroat and rufous-sided towhee preferred areas burned 4 years previously, and few shrub-dwelling birds used understory burns less than 2 years old (Breininger and Smith 1992). Positive correlations between densities of shrub-dwelling birds and mean shrub height suggest that some shrub dwellers would decline under a regime of fire every 7 years or less. However, this decline would not be expected if some patches of habitat remained unburned (Breininger and Schmalzer 1990). Much scrub occurs as patchy mosaic within other vegetation types that have a greater propensity to burn, so burns are naturally patchy.

Frequent Understory Fires

Understory fires occurring at short (5- to 10-year) intervals usually cause minor changes to vegetation composition and structure and likewise to bird communities. Several studies have shown that many bird species resist changes in abundance in frequently underburned forests. Emlen (1970) reports few changes in the bird community during the first 5 months after understory burning in a 20-year-old slash pine forest in Everglades National Park. The fire removed most of the ground cover, dead grass, and litter; defoliated

36

USDA Forest Service Gen. Tech. Rep. RMRS-GTR-42-vol. 1. 2000

most shrubs; and scorched small trees. Trees in the middle and upper canopy were undamaged. Grass and herbs recovered quickly. Over 70 percent of bird species responses were classified as resister, showing little or no change in abundance. No species showed invader responses after the fire. In southeast Arizona ponderosa pine stands, moderately intense prescribed understory fires (with flame lengths up to 4 feet, 1.2 m) consumed nearly half of all snags more than 6 inches (15 cm) dbh, resulting in a net 45 percent decrease in potential nest trees the first year after treatment (Horton and Mannan 1988). Cavity-nesting bird species abundances changed little, however. In contrast to the above studies, a review by Finch and others (1997) reports considerable community change after "cool" understory burns in ponderosa pine. Seed eaters, timber drilling birds, and some aerial insect eaters increased, while timber and foliage gleaners generally decreased.

In the first 2 years after "cool" prescribed understory fires in the Black Hills, the bird community showed mainly resister and exploiter responses (Bock and Bock 1983). Bird abundance in postfire year 1 was nearly twice that in the unburned area, yet in postfire year 2 abundances were similar in burned and unburned areas. Such rapid shifts could not be explained by changes in vegetation structure or composition. Most likely, temporary, rapid increases in food resources attracted bird species to burned areas and resulted in a quick surge in their abundances.

The severity of understory fire affects the composition and abundance of the bird community after fire. In loblolly pine-bottomland hardwood forests of Alabama's Piedmont, high-intensity understory fire removed vegetation from the middle of the canopy down, while low-intensity understory fires had a "patchy" effect, with live and dead understory vegetation interspersed. Significantly more birds used the low-intensity burns than the high-intensity burns in the 4 months after treatment (Barron 1992). Bark, canopy, and shrub gleaners were more abundant on the low-intensity burn, while ground foragers were more abundant on the high-intensity burn.

Infrequent Understory Fires

More substantial changes in forest structure and the bird community may occur after fire in areas with infrequent understory fire (intervals greater than 10 years). Populations of the most common breeding birds decreased after severe understory fires in Yosemite National Park, while less common species increased substantially (Granholm 1982). Two understory fires were examined: a prescribed fire in white fir-mixed conifer forest and a naturally ignited understory fire in a California red fir forest. In presettlement times, these forest types underburned every 17 to 65 years (Taylor and Halpern 1991). Trees up to 40 feet (12 m) tall were killed by the fires. Bird communities in the two burns showed similar responses. The highest proportion of species responses was in the resister category. No species avoided the burns, and more than 70 percent of the responses were classified as resister, exploiter, or invader. Hermit thrush and Hammond's flycatcher populations were reduced most by the fires, and woodpecker populations increased most.

Vegetation usually responds more slowly after fire in dry forests, including pinyon-juniper, than in more productive, frequently burned forests. The bird community may likewise be slow to return to its prefire composition and structure. In pinyon-juniper forests of Nevada, understory fires occurred in the past much less frequently (once every several decades) than in the southern pines and ponderosa pine (Wright and Bailey 1982). In central Nevada, more than 60 percent of bird species the first 2 years after prescribed understory fire showed vacillator (showing wide population fluctuations) or exploiter responses (Mason 1981). Species using resources near the ground increased most after burning. Savannah and black-chinned sparrows were found only on burned areas.

Stand-Replacement Fire Regimes

Research in the literature indicates that bird communities are disrupted for at least 2 years by stand-replacing fire. A few studies show signs that the community is returning to its preburn structure in postfire years 3 and 4, but others do not. The changes can be positive for insect-eating and seed-eating species and negative for species that require a dense, closed canopy such as bark and foliage gleaners.

Grasslands

Grasslands with few or no shrubs have a relatively simple aboveground vegetation structure, which is consumed almost completely by fire. Vegetation change following fire is rapid. Conditions similar to preburn vegetation composition and structure reestablish by postfire year 2 or 3 (for example, see Launchbaugh 1972). Although grasses dominate the vegetation, forbs often increase in density and cover immediately after fire, so plant diversity may be highest within the first 2 years after fire. Bird species that nest and use grasslands seem to be well adapted to rapid, predictable changes in habitat characteristics associated with fires, even though such fires often remove avian nest substrates and hiding cover.

USDA Forest Service Gen. Tech. Rep. RMRS-GTR-42-vol. 1. 2000

37

Bird communities in a South Dakota prairie 2 to 3 months after fire showed dramatic population changes, with a high proportion of invader, endurer, and avoider responses (Huber and Steuter 1984). This was the only grassland study that showed such a high proportion of invader responses, which may be due to the short duration of the study and the fact it was conducted soon after fire. Upland sandpiper and western meadowlark showed substantial increases compared to unburned areas, while grasshopper sparrow and red-winged blackbird had much lower abundances on the burn.

Other research on postfire bird communities was done over longer periods than the above study. During the first 2 years after grassland fires in southeastern Arizona, most bird populations changed, but few species abandoned or were completely new to the area (Bock and Bock 1978). Nearly 75 percent of the species responses were classified as vacillator, endurer, and exploiter.

In Saskatchewan, the bird community also changed in the first 2 years after grassland fire (Zimmerman 1992). More than half of the bird populations showed resister responses. No responses were classified as avoider, and only a few responses were invader and exploiter. Abundance of key species such as clay-colored and savannah sparrows were still substantially below the unburned levels in year 3, so overall abundance was consistently lower in the burned area. Recovery was slower than in other grasslands studied. The cool climate and short growing season of Saskatchewan may slow the recovery process for some prairie species.

The same bird species may respond differently to fire in different habitats. For example, field sparrows in central Illinois prefer to breed in grasslands overgrown with shrubs and young deciduous trees (shrub-grassland), but they also breed in grasslands without brush and in open woodlands (Best 1979). After burning, field sparrows used shrub-grassland more and burned grassland less than they had during the same period the previous year. Thus the response of field sparrow populations in grasslands was endurer, and the response in shrub-grasslands was exploiter. Fire evidently caused field sparrows to use the preferred habitat more intensively than the less-preferred habitat.

Climatic interactions with fire and habitat suitability are not well understood, but adaptation to periodic drought may be essential for a bird species to persist in grass-dominated communities (Zimmerman 1992). In average and wet years, food resources increased in Kansas prairie after fire, yet bird abundance did not. This indicated that the bird community was saturated (Zimmerman 1992). When drought and fire overlapped and resources were reduced, even drought-adapted species decreased in abundance, although no species disappeared from the community.

Shrub-Grasslands

We differentiate between grasslands and shrub-grasslands because grass-dominated areas with shrubs have more complex habitat structure than grasslands. The only shrub-grasslands discussed here are those in which shrubs were present before fire or in unburned areas used as controls. Shrub-grasslands are likely to have more niches available to birds and to recover their preburn structure more slowly after fire than grasslands. The two 1-year studies examined here indicate that annual burning causes substantial changes in bird communities in shrub-grasslands.

Annual burning of a Kansas prairie for more than 10 years led to a significant decrease in bird species richness. Annual burning maintained the prairie with low coverage of woody vegetation, rendering it unsuitable for woody-dependent core species and most other species. Among the bird species present every year, response to fire was almost 90 percent resister, endurer, and avoider (Zimmerman 1992). Annual burning virtually eliminated habitat characteristics needed by Henslow's sparrow and common yellowthroat.

Most species abundances changed in response to fire on a southwestern Florida dry shrub-grassland. In the first postfire year, most species responses on burned plots (with shrub cover ranging from 34 to 82 percent) were invader and avoider, compared to plots without fire for more than 15 years that had a closed shrub canopy (Fitzgerald and Tanner 1992). Species showing an invader response were mostly ground feeders (for example, Bachman's sparrow and common ground-dove), whereas shrub-dwelling species showed the avoider response (for example, northern cardinal and gray catbird). Burned plots provided better avian habitat than mechanically treated plots (in which shrubs were chopped). Birds colonized the burned plots much sooner than the mechanically treated plots. Shrubs killed by fire provided a more complex habitat structure than shrubs in the mechanical treatment. Annual burning would ultimately exclude shrubs, so the bird community response would probably resemble that after mechanical treatment.

The importance of shrubs as perches in shrub-grasslands is illustrated by a study in Kansas tallgrass prairie (Knodel-Montz 1981). Forty artificial perches were placed in burned and unburned prairie. Comparisons were made among plots annually burned and unburned, with and without artificial perches. Artificial perches on the burn were used nearly twice as often as those on the unburned plot, although the difference was not statistically significant. In the unburned plot, birds seemed to prefer natural perches to artificial ones.

38

USDA Forest Service Gen. Tech. Rep. RMRS-GTR-42-vol. 1. 2000

Shrublands

Shrublands usually occur in dry environments and are characterized by sparse to dense shrubs with few or no trees. Examples are the extensive sagebrush lands of the Interior West and California chaparral. Fuels in shrublands tend to burn rapidly. Fires typically move swiftly and are difficult to control. Most aboveground vegetation is consumed by fire, so the structure of the vegetation is altered dramatically. Recovery time ranges from years to decades, depending largely on the resprouting ability of the species burned. Bird populations often decline after shrubland fires, but declines may be offset by populations that rebound if fire spread is patchy, leaving some areas unburned, and if species usually associated with grassland communities invade the burn.

Numbers of bird species and individuals were much lower where fire burned a California coastal sage scrub community dominated by California sagebrush than in an unburned area nearby (Stanton 1986). This was most noticeable the first 18 months after the fire (Moriarty and others 1985). The fire killed all but a few large shrubs and trees. Among the 37 bird species observed by Stanton (excluding raptors), more than 70 percent responded as resister and endurer, and few as avoider or invader. Significant differences in foraging activity between seasons and between burned and unburned areas were observed. All birds except the flycatchers spent more time actively foraging in the unburned than in the burned area. Permanent residents foraged more in the burned area during spring and early summer than during the rest of the year. Birds tended to perch rather than forage in the burned area.

Lower bird populations also predominated after a fire in big sagebrush in south-central Montana that killed nearly 100 percent of the sagebrush (Bock and Bock 1987). In postfire year 3, grass and herb cover were much higher on the burn than in a similar unburned area, but no recruitment of new sagebrush was detected. Of the few species detected, responses were mostly avoider, plus either endurer or resister. Lark sparrow, lark bunting, and Brewer's sparrow all avoided the burn. During the breeding season, these three species occupied patches of significantly more shrub canopy in the unburned area than available randomly. Grasshopper sparrows were classified as endurer. The only resister response was by the western meadowlark, an adaptive grassland bird.

Because sagebrush does not sprout from underground buds after fire, sagebrush communities require several decades to establish postfire vegetation composition and structure similar to that on unburned sites. Incomplete burning, characteristic of sagebrush stands, appears to be important to the development of these communities. Unburned islands of sagebrush are important sources of sagebrush seed after fire and retain habitat features vital to species associated with shrubs, such as sage grouse and Brewer's sparrow.

In southeastern Idaho, more than 50 percent of the species responses were classified as resister for a postburn bird community in big sagebrush (Petersen and Best 1987), where the prescribed fire was incomplete, killing about 50 percent of the shrubs. The first year after fire, total bird abundance declined significantly (22 percent). In years 2 and 3, there were no significant abundance differences between burned and unburned areas. In year 4, significantly more birds were detected on the burn. Species showing resister responses may have used different parts of the patchy postfire habitat. No species avoided the burn during the 4 years of the study.

Nonuniform burning was used to explain bird community changes after a fire in sagebrush in north-central Utah; 3 and 4 years after the fire, few bird species showed appreciable declines. Bird abundance in the burned area (with about 90 percent of aboveground vegetation burned and 80 percent of shrubs killed) was compared to abundance in an unburned site plowed 17 years before the study (Castrale 1982). Total bird density and number of breeding species were similar on the two sites. Breeding bird responses 3 to 4 years after fire were primarily exploiter and resister, with no avoider or invader responses. Burning was associated with increases of western meadowlark, a grassland species. Brewer's sparrow and sage thrasher, which nest above the ground in shrubs, were associated with unburned islands of sagebrush and did not use the grassy portions of the burned site. If the fire had killed all the shrubs, Brewer's sparrow, which can be eliminated from sagebrush habitats with chemical control of shrubs (Schroeder and Sturges 1975), probably would have been absent.

Completeness of burn influenced fire effects in central Florida's oak scrub (Breininger and Schmalzer 1990). During the winter and spring following a November burn, stations with more than 95 percent of the vegetation burned had low numbers of permanent residents, while stations with 10 to 25 percent of the vegetation burned had bird counts similar to unburned stations.

The nature of habitat adjacent to burned shrublands sometimes influences bird community responses. Lawrence (1966) sampled bird communities in interior California chaparral dominated by buckbrush before a prescribed fire and 3 years afterward. Observation transects crossed chaparral, grassland, and pine-oak woodland. Most chaparral species responses were classified as resister and endurer, with no avoider or exploiter responses. California quail and scrub jay declined sharply after fire.

USDA Forest Service Gen. Tech. Rep. RMRS-GTR-42-vol. 1. 2000

39

Research in California shrublands indicates that fire does not reduce species diversity but does alter species composition. During the year following a stand-replacing fire in coastal sage scrub, southern California, the species richness of birds in the burned area gradually increased. By the end of the first year, species richness on the burn was 70 to 90 percent similar to that on an adjacent unburned area (Moriarty and others 1985). The species most abundant in the burn were those typically associated with open areas, whereas the species most abundant in unburned areas typically avoid open areas.

Forests and Woodlands

Stand-replacing fires in forests and woodlands are either severe surface fires or crown fires; more than 80 percent of the trees are top-killed or killed. The contrasts between prefire and postfire environments are much sharper than after understory fire, and the time needed for the vegetation to develop structure and composition resembling the preburn forest is measured in decades to centuries. During this time, many forces can alter the trajectory of succession, so the mature forest may differ substantially from the preburn forest. A stand-replacing fire is likely to result in many or most of the bird species present before fire being replaced by new species (Finch and others 1997). Some species use habitat that occurs only for a short time after stand-replacing fire. In Yellowstone and Grand Teton National Parks, more species were unique to the postfire period (1 to 17 years) than to later stages of succession (111 to 304 years after stand-replacing fire) (Taylor and Barmore 1980).

In this section we first describe bird response to fire in the short term (less than 5 years after fire) and then in the long-term (5 years or longer). Short-term studies typically included control plots, either sampled before the fire or after the fire in a similar, unburned area. Long-term studies covered early to late stages of vegetation succession. Some examined succession from 6 to 60 years after fire, when canopy closure occurred. Others examined a chronosequence of similar sites at different locations from early to late seral conditions.

Short term—The few studies available indicate that changes in habitat characteristics caused by stand-replacing fire cause postfire avian communities to differ substantially in the short term from prefire communities. High turnover occurs in the first 5 years after stand-replacing fire. The predominant response categories are invader and avoider. These responses usually describe 50 to 90 percent of postfire bird populations. Few species responses are classified as resister after crown fire, often less than 20 percent of the species present in the first 2 years postfire; some

studies show no resister responses to fire. This community response to fire differs substantially from the response generally observed in understory fire regime types, where a high proportion of the postfire bird community consists of resister species. Most studies of understory fire regimes showed at least a third of the species responses as resister, with some over 70 percent.

In western hemlock forests of western Washington, which has a stand-replacing fire return interval spanning several centuries, more than half the bird populations showed invader and avoider responses during the first 2 years after a severe crown fire. The bird community composition shifted from domination by canopy-dwelling species to species nesting and foraging near the ground (Huff and others 1985).

Bird community response to stand-replacing fire in ponderosa pine forests of Arizona was similar to that in western hemlock forests (Lowe and others 1978), even though the climate and presettlement fire regimes of the two communities differ. Nearly 60 percent of the species responses were classified as invader and avoider 1 year after fire.

Substantial species turnover also characterized a dense 200-year-old spruce-fir-lodgepole pine forest in Grand Teton National Park, Wyoming, which burned in stand-replacing fire. More than 80 percent of bird population responses were avoider and invader during the first 3 years postfire (Taylor and Barmore 1980). Few species showed resister responses. As in western Washington, a shift in the bird community from canopy dwellers to ground/brush dwellers occurred. Patterns observed nearby in Yellowstone National Park were similar (Pfister 1980). In 250-year-old lodgepole pine-spruce-fir forest, about three-fourths of the bird community responses were classified as invader in years 2 to 3 after stand-replacing fire. The increased bird diversity in comparison with unburned forest was associated with rapid changes in forest structure and composition after the fire, which attracted several species uncharacteristic of the unburned forest.

A shift from canopy dwelling to ground- and shrub-dwelling species also occurred after stand-replacing fire in northern Minnesota. Apfelbaum and Haney (1981) sampled birds before and after crown fire in a 73-year-old jack pine/black spruce forest. The fire burned severely in an upland pine-dominated area while only lightly burning the hardwood draws. The number of breeding territories decreased by more than half the first year after fire. Tree canopy dwellers were most abundant before the fire, while ground- and shrub-dwelling species predominated afterward. The bird community showed high species turnover; 70 percent of species responses were avoider and invader. The black-backed woodpecker was an

40

USDA Forest Service Gen. Tech. Rep. RMRS-GTR-42-vol. 1. 2000

important species showing the invader response, comprising about 13 percent of total bird abundance after the fire. Ovenbird, the most important ground- and brush-dwelling species prior to fire, avoided the burned area, where the moss ground cover was replaced by lush herbs and jack pine seedlings.

Long term—Oliver and others (1998) show how a "landscape" disturbance is likely to affect bird abundance in three groups of species: those that reside in structurally complex old-growth stands with abundant understory, those that prefer edges between dense and open vegetation, and those that prefer open habitat (fig. 14). The diagram reflects some patterns reported in long-term studies of birds in forested ecosystems, although it does not account for the complex role of fire in producing and destroying snags (see "Snags and Dead Wood" in chapter 1). The predictions are in agreement with Finch and others' (1997) review of the general pattern of species change in southwestern ponderosa pine forests, whether burned by understory or stand-replacing fire: Granivores, tree drilling birds, and some aerial insectivores usually increase after fires, while tree- and foliage-gleaning birds generally decrease. Birds more closely tied to foliage availability, such as hermit thrush and blue-headed vireo, begin recovering as foliage volume increases in subsequent years. Finch and others (1997) add that woodpecker abundance often peaks in the first decade after fire, then gradually declines.

Figure 14 depicts a period early in succession after stand-replacement fire when birds are abundant, and also a time of transition when dominance by open- and

edge-using species gives way to dominance by understory- and canopy-using species. Three studies of bird community dynamics after stand-replacing fire provide some insight regarding the species and habitat requirements that account for these changes. Research on bird community response to succession in the long-term requires either commitment to a long-term research program or use of a chronosequence, a series of sites similar in all characteristics except time since fire. The former method was used for a study in the California Sierra Nevada (Bock and Lynch 1970; Bock and others 1978; Raphael and others 1987). The two other studies discussed here are based on chronosequences. These three studies indicate that (1) early seral conditions foster high bird diversity, (2) more bird species breed exclusively in early seral stages than in mature forests, and (3) snags are a key habitat feature for avian diversity and abundance.

In the Sierra Nevada study, burned and unburned plots were established in 1966, 6 years after a large (approximately 37,000 acres, 15,000 ha) stand-replacing fire in a mixed-conifer forest dominated by Jeffrey pine and white fir. The fire killed nearly all the overstory and understory trees, although small pockets of trees were alive in postfire year 6. Birds were sampled every year except one for the next 20 years.

Changes in the avian community in the burn were primarily related to changes in vegetation structure with succession (Raphael and others 1987). In postfire years 6 to 8, bird abundance on burned plots was similar to that on unburned plots, but species composition differed. Species nesting and foraging on living trees were most abundant on unburned plots, while species characteristic of low brush and open ground predominated in the burned area (Bock and Lynch 1970). Primary cavity excavators (woodpeckers) were more abundant on the burn; even higher numbers may have been present during the 6 years before the study was initiated. Of 32 regularly breeding species, 28 percent were unique to the burn, while 19 percent were unique to the unburned area.

Bird diversity decreased in the burn from postfire years 8 to 15, to less than in the unburned area (Bock and others 1978). By postfire year 15, fewer dead snags were standing and the ground cover was more dense, resembling a shrubfield. Six of 11 hole-nesting species declined during this period. Species that require some open ground, such as dark-eyed juncos—the most abundant breeding species in postfire year 8—were replaced by species indicative of shrubfields, including fox sparrows.

Shrub cover doubled between postfire years 8 and 23, snag density declined 90 percent, and cover of herb and grass-like species decreased significantly. At the end of this period, large snags (more than 15 inches, 38 cm dbh) were 2.5 times more abundant in the

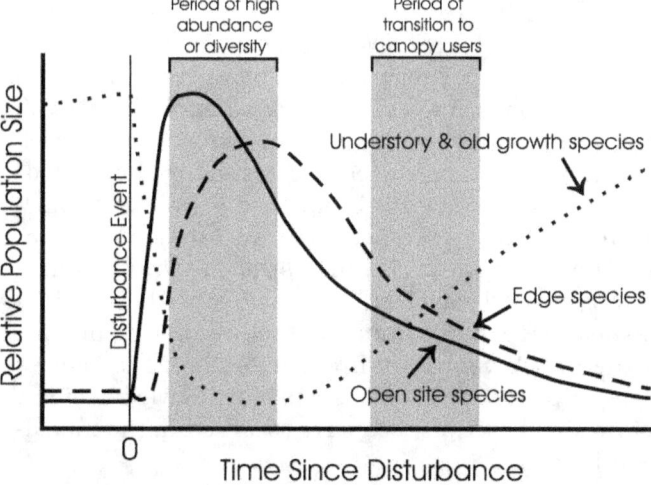

Figure 14—Hypothetical patterns of change in populations of species dependent on three features of forest structure: dense understory/old growth, edge, and open sites. Shaded areas are discussed in the text. Adapted from Oliver and others (1998).

USDA Forest Service Gen. Tech. Rep. RMRS-GTR-42-vol. 1. 2000

41

unburned area than in the burn. Birds that feed and nest in shrubs increased in abundance by more than 500 percent. Woodpeckers declined steadily. At the end of the 25 years, woodpecker abundance on the burn was similar to that in the unburned area. At postfire year 25, vegetation characteristic of a closed-canopy forest still had not developed in the burn. The transition from open- to closed-site species (postulated in fig. 14) was beginning, but the bird community was likely to continue changing and not likely to closely resemble either the unburned area or the burn anytime soon.

Western hemlock forests of different ages (times since stand-replacing fire) were sampled in western Washington (Huff 1984; Huff and others 1985). The ages of stands in this chronosequence were 1 to 3 years, 19, 110, 181, and 515 years. Year 19 of the sere had the highest bird diversity and least resembled other successional stages examined. Lowest bird diversity and abundance occurred at the 110-year-old site where, comparatively, the tree vertical structure was simple, snag density low, and understory composition and structure poorly developed. This stand age may represent the transition from open to closed structure depicted in figure 14. Huff and others (1985) note that rate of forest reestablishment may be slower in western hemlock forests than in the Jeffrey pine-white fir forests of the Sierra Nevada (described above). If so, a longer period of high diversity associated with early seral conditions can be expected for the western Washington sere. Once a full canopy develops in the western hemlock sere, few changes occur in bird species composition. Because the fire return interval is long, species composition may change relatively little for centuries.

A large-scale examination of avian successional relationships after crown fire was conducted in Yellowstone and Grand Teton National Parks, Wyoming, by comparing recent burns to older burns and to areas unburned for at least 300 years (Taylor 1969, 1979; Taylor and Barmore 1980). The most obvious changes in species composition the first few years after fire were surges in abundance of black-backed and northern three-toed woodpeckers. (Prior to the 1974 Waterfall Canyon fire, the black-backed woodpecker was not even known to occur in the Grand Tetons.) Breeding bird density in postfire years 5 to 29 was more than 50 percent greater than in lodgepole pine stands more than 40 years old with closed canopy. In postfire years 5 to 25, following an influx of cavity

excavators, the number of secondary cavity nesters increased rapidly. Two of these species, the tree swallow and the mountain bluebird, dominated the avifauna. They consistently comprised 30 percent or more of postfire birds during the first 30 years after fire. In the second decade after fire, they comprised 55 to 64 percent of the total bird population. By about postfire year 30, mountain bluebirds and tree swallows started to decline at a rate that depended on the loss of standing snags with nest cavities. During this period, vegetation structure and succession made a transition from shrubland to young forest.

The most important event in succession for the postfire bird community was the transition from open to closed canopy, which occurred between postfire years 30 and 50. With the onset of this event, species abundance decreased by more than 60 percent. Species characteristic of later seral stages gradually appeared as the trees got taller. From about postfire year 50 to year 100, change in forest composition and structure stagnated. Over the next 200 years, lodgepole pine in the canopy gave way to shade-tolerant spruce and fir. The bird community changed little during this 250 years, with bird abundance lower than that in earlier successional stages. Bird density and diversity in 300-year-old and older spruce-fir forest is higher than in the previous 250 years.

Mixed-Severity Fire Regimes

Little is known about the effects of fire on bird populations in mixed-severity fire regimes. One might expect the bird community response to mixed-severity fire to be intermediate between responses to understory and stand-replacement fire. Both mixed-severity and stand-replacement fire occurred in Grand Teton National Park, Wyoming, in a 250 year old spruce-fir forest (Taylor and Barmore 1980). Half the species responses were invader and exploiter for the first 3 years after fire. Some canopy-dwelling species typical of unburned areas occurred in the mixed-severity burn but were absent from the stand-replacing burn. These included western tanager, golden-crowned kinglet, red-breasted nuthatch, and yellow-rumped warbler. The mixed-severity burn had less species turnover than the stand-replacement burn in the first 2 years postfire. Almost half the species responses to the stand-replacement fires were avoider, yet no avoider responses were recorded in the mixed-severity burns.

L. Jack Lyon
Mark H. Huff
Jane Kapler Smith

Chapter 6:
Fire Effects on Fauna at Landscape Scales

Studies of disturbance and succession have been a major focus of ecology over the past century (McIntosh 1985). These are studies of temporal pattern. The study of the spatial patterns associated with temporal patterns has blossomed only since about 1990 (Turner 1990). At the landscape scale (25,000 acres, 10,000 ha or more), a complex web of interactions and relationships unfolds. Interactions at this scale are widely accepted as important aspects of ecosystems (see, for example, Agee 1998; Lerzman and others 1998). However, knowledge gained at finer scales of resolution (for example, stand or homogeneous patch) is often difficult to apply at a landscape scale (Schmoldt and others 1999). Including landscape considerations in management demands new approaches to planning, analysis, and design (Diaz and Apostol 1992).

Landscapes are spatially heterogeneous, characterized by structure, function, and temporal variation (Forman and Godron 1986). Landscape structure encompasses the spatial characteristics of biotic and abiotic components in an area and is described by the arrangement, size, shape, number, and kind of patches (homogeneous units). Landscape function is defined by interactions among biotic and abiotic components. Temporal variation of a landscape is expressed by changes in structure and function over time. Configuration of patches affects the occurrence and spread of subsequent fires, so landscape-level feedback is an important part of fire effects at landscape scales (Agee 1998).

Fire's most obvious function in landscapes is to create and maintain a mosaic of different kinds of vegetation (Mushinsky and Gibson 1991). This includes size, composition, and structure of patches, as well as connectivity (linkages and flows) among patches. Within a large (200 mi^2, 500 km^2) burn in Alaska, Gasaway and DuBois (1985) reported substantial variation in fire severity and many unburned patches, resulting in variation in plant mortality and perpetuation of the mosaic nature of the landscape. Over time, a mixture of a few large burns with many small burns and variation within them produces relatively small homogeneous areas. One study in northern Manitoba reported an average stand size of 10 acres (4 ha) (Miller 1976 in Telfer 1993). Stand-replacing fires in boreal forest may skip as much as 15 to 20 percent of the area within their perimeters. The 1988 fires in the Greater Yellowstone Area, well publicized because of their size and severe fire behavior, actually consisted of a complex patchwork containing areas burned by crown fire, areas burned by

USDA Forest Service Gen. Tech. Rep. RMRS-GTR-42-vol. 1. 2000

43

Figure 15—Aerial photo shows variation in fire severity over the landscape after the 1988 fires in the Greater Yellowstone Ecosystem. Black patches were burned by crown fire. Most of these are surrounded by red and gray areas where trees were killed by severe surface fire. Green cover represents a combination of unburned forest and areas burned by understory fire. Photo by Jim Peaco, courtesy of National Park Service.

severe surface fire, underburned sites, and unburned areas (Rothermel and others 1994) (fig. 15, table 4). The majority of severely burned area was within 650 feet (200 m) of unburned or "lightly burned" areas.

Landscape-scale fire effects on fauna include (1) changes in availability of habitat patches and heterogeneity within them, (2) changes in the composition and structure of larger areas, such as watersheds, which provide the spatial context for habitat patches, and (3) changes in connections among habitat patches. During the course of postfire succession, all three of these landscape features are in flux.

Fire changes the proportions and arrangement of habitat patches on the landscape. When fire increases heterogeneity on the landscape, animal species have increased opportunities to select from a variety of habitat conditions and successional stages. Fires often burn with varying severity, increasing heterogeneity. Adjacent unburned areas (which may surround or be embedded in the burn) serve as both sinks and sources for animal populations, and also influence animal

emigration and immigration patterns (see Pulliam 1988). Bird diversity after stand-replacing fire may be higher on patchy or small burns than on large, uniform burns because the small areas are accessible to canopy and edge species as well as species that use open areas. A small (300 acre, 122 ha) stand-replacing fire in Douglas-fir forest in western Montana

Table 4—Proportion of area burned at four severities within the perimeter burned each day in the Greater Yellowstone Area, 1988 (Turner and others 1994).

Severity level	Percent of area burned daily	
	June 1-July 31	Aug. 20-Sept. 15
Unburned	29.3	28.2
Underburned, "light" burn	18.9	14.5
Severe (stand-replacing) surface fire	26.6	24.4
Crown fire	25.1	32.8

44

USDA Forest Service Gen. Tech. Rep. RMRS-GTR-42-vol. 1. 2000

left many unburned patches. The burn attracted wood-boring insects, woodpeckers, and warblers. The burn itself was not used by Swainson's thrushes, but they remained abundant in nearby unburned areas (Lyon and Marzluff 1985).

Two management examples show how understanding of the relationship of individual species to landscape heterogeneity can be applied. The Karner blue butterfly (fig. 16) requires wild lupine, a forb growing in fire-dependent oak savanna and prairie, to complete its life cycle. The larva itself (fig. 17), however, is very sensitive to fire. To protect the butterfly at Indiana Dunes National Lakeshore, managers divide the landscape so that every burn area contains patches from which fire is excluded; these patches serve as refugia from which the butterfly can repopulate the burn (Kwilosz and Knutson 1999).

The sage grouse is sensitive to fire effects on the arrangement of habitat components on the landscape. Stand-replacing fire in sagebrush changes the proportions and arrangement of sage grouse habitat components. It is this arrangement that determines whether fire benefits or damages the species. Sage grouse use various successional stages of the sagebrush sere as lekking, nesting, brooding, and wintering grounds. Forb and insect availability are the driving factors in sage grouse productivity (Drut and others 1994). Fires increase openings, which often increases forb production. Fires may also enhance the nutritional value of browse and provide new lekking sites (Benson and others 1991; Martin 1990; Pyle and Crawford 1996). If burns cover large tracts of

sagebrush or remove sagebrush from key wintering areas, however, they may damage sage grouse populations (Fischer and others 1996; Gregg and others 1994; Klebenow 1969, 1973; Welch and others 1990). Neither extensive dense sagebrush nor extensive open areas constitute optimal habitat for the species. While burning sometimes succeeds in restoring the balance of plant community components in sage grouse habitat, it is accompanied by the risk of increasing cheatgrass productivity, which may cause the area to reburn before sagebrush recovers (Crawford 1999).

Researchers in many ecosystems recommend addressing the size and spatial arrangement of patches in planning for specific objectives. In Southeastern forests, Dunaway (1976) recommends interspersion of underburned areas in longleaf pine, which have low ground cover and provide successful foraging for northern bobwhite chicks, with unburned areas for

Figure 17—Karner blue butterfly larva feeding on its sole food source, the fire-dependent wild lupine. Ants protect the larvae from predation and feed on "honeydew," a high-sugar liquid exuded by the larvae. Photo by Catherine Papp Herms, courtesy of the Michigan Chapter, The Nature Conservancy.

Figure 16—Karner blue butterfly, an endangered species whose larval form feeds exclusively on a the fire-dependent wild lupine. Photo by Robert Carr, courtesy of the Michigan Chapter, The Nature Conservancy.

USDA Forest Service Gen. Tech. Rep. RMRS-GTR-42-vol. 1. 2000

45

escape cover and sheltering broods. In the Western States, Belsky (1996) suggests that a mosaic of pinyon-juniper woodland, grassland, and intermediate seral communities would optimize biodiversity in arid Western ecosystems. In the Southwest, Reynolds and others (1992) quantify the proportions of a landscape in ponderosa pine forests, characterized in presettlement times by understory fire regimes, that seem desirable for sustaining northern goshawk populations (fig. 18). Recommendations for sustaining habitat and prey for the northern goshawk in Utah and the Rocky Mountains include increasing the predominance of early-seral and midseral species, increasing the numbers of large trees in the landscape, and maintaining connectivity among habitat patches (Graham and others 1997, 1999).

Some animals require habitat that contains different features at different scales. Wright (1996) found many patches of old-growth ponderosa pine and Douglas-fir in western Montana that seemed suitable for occupation by flammulated owls, but the owls occupied fewer than half of them. The explanation lies in the landscape context for the patches of old growth. Occupied patches (fig. 19) were embedded in a landscape with many grassy openings and some dense thickets of Douglas-fir; unoccupied patches (fig. 20) were typically embedded in a landscape of closed, mature forest. Understory fire may enhance old growth for nesting, openings for foraging, and the landscape context for nest sites. However, a homogeneous underburned

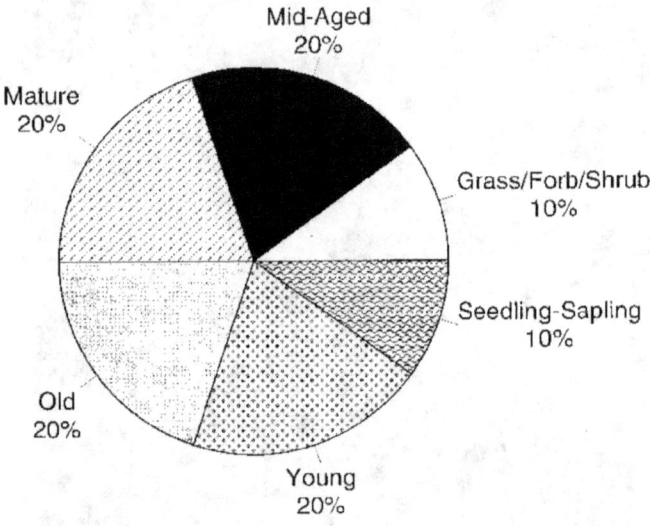

Figure 18—Proportions of a home range or landscape desirable for sustaining goshawks in forests with understory fire regimes (Reynolds and others 1992), example of a set of recommended stand descriptors to be implemented at landscape scales.

landscape without Douglas-fir thickets would reduce the quality of habitat for the owl.

Nesting patterns of the Florida scrub-jay provide a second example of habitat requirements that vary according to scale. Optimum habitat for Florida scrub-jays consists of open oak patches 10 years or more after fire containing openings that often result from more recent fires. Scrub-jays use these openings for caching acorns (Breininger and others 1995). The oak patches preferred for nesting occur within a matrix of pine-scrub habitat, which is not used directly by the jay but indirectly serves its needs by providing prey species, enabling jays to see predatory birds from a long distance, and spreading fires into oak-dominated areas, which often burn poorly. Management that favors open oak without considering the more flammable adjacent habitat can result in a loss of openings and an increase in shrub height and tree densities, and eventually a Florida scrub-jay decline (Breininger and others 1995).

Corridors and connectivity influence habitat use by migratory fauna such as bison (Campbell and Hinkes 1983) and caribou (Thomas and others 1995), and for many predators, including fisher (Powell and Zielinski 1994), lynx (Koehler and Aubry 1994) and spotted owl (Laymon 1985; Thomas and others 1990). Connectivity is a crucial consideration for aquatic fauna as well (Rieman and others 1997). Although research design considerations may make it difficult to demonstrate conclusively that wildlife corridors benefit fauna (Beier and Noss 1998), some species definitely require landscapes with little fragmentation and high connectivity (Bunnell 1995). Bighorn sheep in Alberta foraged in spruce-pine forest burned 10 years previously by stand-replacing fire significantly more than in unburned areas, probably because the burned sites had a more open structure adjacent to escape areas (Bentz and Woodard 1988). Connectivity accounted in part for the expansion of a bison herd in Alaska after fire. Stand-replacing fire in black spruce forest produced extensive sedge-grasslands, a type that bison depend upon for winter range (Campbell and Hinkes 1983). The authors comment that winter range expansion was enhanced because the burned area was contiguous with summer range and areas used for winter range prior to the fire, so access to burned range was relatively easy. Where black spruce and shrublands fragment sedge-grasslands, bison have difficulty accessing their winter range because of deep snowpack.

In landscapes that contained a fine-grained mosaic of structures and age classes during presettlement times, native fauna could readily find most kinds of habitat. In contrast, in landscapes where large, stand-replacing fires were common, fauna sometimes traveled great distances in search of habitat. Ecosystems with large, stand-replacing fire regimes and

Figure 19—Not all habitat that seems suitable for flammulated owls at the stand level is occupied by the owl; suitability at the landscape is also important. This photo depicts a typical landscape in western Montana where flammulated owls *were* detected. Photo by Vita Wright.

Figure 20—Typical landscape where flammulated owls were *not* detected, western Montana. Photo by Vita Wright.

USDA Forest Service Gen. Tech. Rep. RMRS-GTR-42-vol. 1. 2000

47

ecosystems that are now highly fragmented probably require more attention to connectivity than areas retaining a fine-grained mosaic. As fire and fire exclusion further alter landscapes, corridors, and entry/exit areas near corridors, fauna that require large, continuous areas of any structure—whether early seral or old growth—may not readily find new habitat.

Despite the tendency of natural fire regimes to provide habitat with a variety of structures at a variety of successional stages, one cannot assume that landscapes prior to European American settlement were at equilibrium, even at a landscape scale (Agee 1998). Ecosystems characterized in past centuries by infrequent large, severe fire are especially unlikely to exhibit a steady-state age structure because large fires have a long-term effect on the distribution of age classes on the landscape. Examples include aspen-black spruce forests in Alberta (Cumming and others 1996) and lodgepole pine in the Greater Yellowstone Area (Turner and others 1994). Presettlement fire regimes are an important part of the context for management. The spatial and temporal variability in these regimes, though difficult to identify and apply, may be a crucial aspect of effective management (Lertzman and others 1998).

Effects of Altered Fire Regimes ____

Excluding fire from a landscape, unless it is being intensively managed for fiber production, has two major effects on animal habitat. First, it increases the abundance and continuity of late-successional stages. Second, it changes fuel quantities and fuel arrangement, at least for a time.

Extensive changes in habitat associated with decades of fire exclusion are most evident in areas influenced by frequent fires during presettlement times (Gruell and others 1982). Understory fire regimes in southeastern forests provide one example. Fire exclusion increases dominance by less flammable vegetation, converting pine to hardwood forest (Engstrom and others 1984). Loss of early seral structures on sandy sites contributes to the decline of several reptile species, including sand skinks, six-lined racerunners, mole skinks, and central Florida crowned snakes (Russell and others 1999).

Kirtland's warbler/jack pine ecology in Michigan provides another example of fire exclusion's effects at landscape scales. Jack pine forests were characterized in presettlement times by relatively frequent, stand-replacement fire. The Kirtland's warbler (fig. 21)

Figure 21—Kirtland's warblers at nest. Photo by Betty Cotrille, courtesy of the Michigan Chapter, The Nature Conservancy.

nests on the ground in dense jack pine regeneration 5 to 24 years after stand-replacing fire or harvesting (Mayfield 1963; Probst and Weinrich 1993). The warbler was nearly extirpated during the 1960s and 1970s because of nest parasitism by the brown-headed cowbird, fire exclusion, and tree regeneration practices in jack pine forests of Michigan (Mayfield 1963, 1993). Extensive use of fire and harvesting to provide breeding habitat have kept the Kirtland's warbler from extinction, although uncertainty still exists about habitat attributes that actually limit population growth (nest sites, lower branch cover for fledglings, and foliage volume for foraging). Habitat modeling and management planning need to integrate habitat requirements, dynamics of disturbance and succession over large areas, and population dynamics of the warbler itself (Probst and Weinrich 1993).

In many Western ecosystems, landscape changes due to fire exclusion have changed fuel quantities and arrangement, increasing the likelihood of increased fire size and severity (Lehman and Allendorf 1989). In interior ponderosa pine/Douglas-fir forest, for example, exclusion of understory fire has led to the development of landscapes with extensive ladder fuels, nearly continuous thickets of dense tree regeneration, and large areas of late successional forest infested with root disease (Mutch and others 1993). These changes not only constitute habitat loss for species that require open old-growth stands and early seral stages; they also may increase the likelihood of large, severe fires in the future. Where fire exclusion has caused a shift in species composition and fuel arrays over large areas, subsequent fires without prior fuel modification are not likely to restore presettlement vegetation and habitat (Agee 1998).

The effects of fire exclusion on fauna that require late-seral and old-growth habitat originally established by fire are largely unknown. Although pileated woodpeckers do not nest in recent stand-replacement burns, they do prefer to nest in western larch, a fire dependent tree, in the Northern Rocky Mountains (McClelland 1977). If altered fire regimes reduce the abundance of large, old western larch, they are likely to impact the woodpecker as well. In presettlement times, the spotted owl occupied landscapes that consisted of large areas of forest at different stages of succession, characterized by Gaines and others (1997) as a "very dynamic" landscape. The owl prefers old-growth forest within this landscape, so fire exclusion has enhanced owl habitat, at least in some parts of the owl's range (Thomas and others 1990). Large, severe fires now would reduce the species' habitat and reduce connectivity between remaining old-growth stands (Thomas and others 1990). Protection of the owl may include fuel reduction in areas adjacent to occupied stands to reduce the likelihood of stand-replacement fire.

Several ecosystems in Western North America experience more frequent fire now than they did in the past because of invasive species. Where cheatgrass dominates areas formerly covered by large patches of sagebrush and grassland, for example, fires now occur almost annually and shrub cover is declining. Knick and Rotenberry (1995) report that site selection by sage sparrow, Brewer's sparrow, and sage thrasher is positively correlated with sagebrush cover. In addition, sage sparrow and sage thrasher prefer large to small patches of shrubs. The sage grouse requires mature sagebrush as part of its habitat, so extensive stand-replacing burns are likely to reduce its populations (Benson and others 1991). Increased fire frequency and cheatgrass cover have increased landscape-level heterogeneity by reducing sagebrush cover and patch size, lowering the value of even remnant sagebrush patches as habitat for native birds (Knick 1999).

USDA Forest Service Gen. Tech. Rep. RMRS-GTR-42-vol. 1. 2000

49

Notes

_____

_____

_____

_____

_____

_____

L. Jack Lyon
Robert G. Hooper
Edmund S. Telfer
David Scott Schreiner

Chapter 7:
Fire Effects on Wildlife Foods

Fire's influence on wildlife food is probably the most thoroughly researched aspect of the relationship between fire and fauna. It has generated a vast literature showing a great variety of results. The literature is not balanced among faunal classes or geographic regions, since most studies focus on a species of concern to managers in a particular geographic area. To summarize this literature, we return to the vegetation communities described in chapter 2, organizing the discussion according to the five geographic regions presented there. This structure also corresponds to the broad outline of "Effects of Fire on Flora" in the Rainbow Series.

For each vegetation type, we summarize what is known about changes in quantity and quality of forage following fire. In addition, where information is available, we summarize changes in availability of seeds, mast, and insects.

The factors that affect postfire changes in vegetation quantity and nutritional quality include soil, vegetation type, age and structure of vegetation prior to burning, rainfall before and after burning, severity of the fire, season of burning, time since fire, and pre-settlement disturbance regime. In general, the literature regarding fire effects on wildlife food indicate that:

- Burning sets back plant development and succession, often increasing or improving forage for wildlife from a few years to more than 100 years, depending on vegetation type.
- Fires usually increase habitat patchiness, providing wildlife with a diversity of vegetation conditions from which to select food and cover.
- The biomass of forage plants usually increases after burning in all but dry ecosystems.
- The production of seeds by grasses and legumes is usually enhanced by annual or biennial fires. Mast production is usually enhanced by a 5-year or longer burning cycle.
- Burning sometimes, but not always, increases the nutritional content and digestibility of plants. This effect is short-lived, typically lasting only one or two growing seasons.
- Some wildlife species select a more nutritious diet from burned areas even though the average nutrient content of burned plants does not differ from that of unburned plants.

It is impossible to generalize about fire effects on wildlife foods that apply throughout all of North America. Furthermore, while improving and increasing food for particular wildlife species may be an

USDA Forest Service Gen. Tech. Rep. RMRS-GTR-42-vol. 1. 2000

51

important objective, it is important that these goals not be accomplished at the expense of ecosystem sustainability (Provencher and others 1998).

Northern Ecosystems

Boreal Forest

Fire and the Quantity of Forage—During the first two growing seasons after a mixed-severity fire with large areas of stand-replacement in Minnesota boreal forest, most herbaceous and low shrub species increased rapidly in biomass. Production leveled off during the next 3 years (Irwin 1985; Ohmann and Grigal 1979).

Stand-replacing fire in boreal forest can greatly increase the production of woody browse for moose (MacCracken and Viereck 1990; Oldemeyer and others 1977; Wolff 1978). Prefire stand age and species composition play a significant role in plant response to fire (Auclair 1983; Furyaev and others 1983; MacCracken and Viereck 1990; Viereck 1983). Aspen stands that were 70 years old before stand-replacing fire produced 10 times as much browse in the first postburn year than did birch and spruce stands that were 180 years old before fire. Spruce stands that were 70 years old before fire produced three times as much browse after burning than did similar stands 180 years old before burning. The benefits of burning to moose may peak 20 to 25 years after stand-replacing fire (MacCracken and Viereck 1990; Oldemeyer and others 1977) and last less than 50 years (Schwartz and Franzmann 1989).

In boreal forests, stand-replacing fire reduces the lichens that caribou use as forage in winter; lichens may be reduced for up to 50 years after fire. Caribou prefer open forests burned 150 to 250 years ago. Their preference is related not only to abundance of food, but also to snow cover, visibility of predators and other herd members, and nearness to traditional travel routes (Thomas and others 1995). Lichens decline in old stands (200 years or more), indicating that fires of moderate to high severity may be essential for maintaining forage for caribou in the long-term (Auclair 1983; Klein 1982; Schaefer and Pruitt 1991).

Fire and Nutritional Quality—Stand-replacing fire in boreal forest increases the protein, phosphorus, calcium, magnesium, and potassium content of woody browse for moose, but this effect is probably gone by the third growing season (MacCracken and Viereck 1990; Oldemeyer and others 1977).

Laurentian Forest

Fire and the Quantity of Forage—Quaking aspen and paper birch, two of the most important browse plants for white-tailed deer and moose in the Northern and Eastern States, both sprout well after most fires. Paper birch reaches peak browse production 10 to 16 years after stand-replacing fire (Safford and others 1990). Twenty-five years after prescribed fire in quaking aspen in northern Minnesota, aspen productivity was 111 percent of productivity on unburned stands (Perala 1995).

Fire and Nutritional Quality—Low-intensity understory fires in aspen stands in southern Ontario increased the levels of nitrogen, calcium, phosphorus, magnesium, and potassium in aspen leaves the first growing season after burning (James and Smith 1977). The level of potassium in twigs was lower in burned than unburned stands.

Increases in some nutrients have been reported after severe fire. Ohmann and Grigal (1979) reported the effects of a mixed-severity fire (with large areas of stand-replacement) in forests of jack pine, quaking aspen, and paper birch in northern Minnesota. Concentrations of potassium, calcium, and magnesium increased during the first 5 years after fire, generally exceeding levels on unburned sites. Phosphorus on burned sites also exceeded that on unburned sites. Nitrogen concentrations were higher on burned than unburned sites but declined during the first five growing seasons after fire.

Eastern Ecosystems and the Great Plains

Eastern Deciduous Forests

Fire and the Quantity of Forage—Biomass of herbs and shrubs usually increases after fire in Eastern deciduous forests. The fire frequency needed for maximum productivity differs among vegetation types. Prescribed understory burns at 10- and 15-year intervals did not affect shrub or herbaceous cover in New Jersey pine stands in comparison with unburned sites (Buell and Cantlon 1953). As intervals between burns decreased to 5, 4, 3, 2, and 1 years, however, the shrub cover decreased and cover of herbaceous plants, mosses and lichens increased. Two and three growing seasons after late winter-early spring prescribed burns in oak-hickory stands in West Virginia (Pack and others 1988), herbaceous vegetation increased, although response to burning varied considerably. Results suggested that thinning stands prior to prescribed burning was necessary to increase understory productivity.

Fire and Nutritional Quality—Many studies from the Northeast and Midwest report increased nutrition of wildlife foods after fire, but the duration of increases varies. Prescribed understory burns in April increased the July levels of crude protein in

52

USDA Forest Service Gen. Tech. Rep. RMRS-GTR-42-vol. 1. 2000

scrub oak foliage in central Pennsylvania for 4 years (Hallisey and Wood 1976). Crude protein also increased in blueberry foliage, but only during the first growing season. Phosphorus, magnesium, and calcium levels were higher in scrub oak foliage, and magnesium was higher in teaberry the first growing season after fire. During the growing season after an April understory burn in mixed-oak forests in Wisconsin, the concentration of nitrogen, phosphorus, and potassium in leaves increased (Reich and others 1990). The increase was thought to be due to increased availability of nutrients in the soil. For most nutrients in most plant species, the effect decreased throughout the growing season.

Southeastern Forests

Fire and the Quantity of Forage—Several studies indicate that, in general, understory fire in Southern forests does not increase the biomass of forage but often increases the proportion of herbage to browse (Evans and others 1991; Stransky and Harlow 1981; White and others 1991) (fig. 22). Other studies have found that biomass increased after burning only under some conditions. Gilliam (1991) reported an increase in herbaceous biomass after burning a

pine-bluestem range in Louisiana that had not been burned in 40 years. Prescribed burning combined with herbicides significantly increased the amount of forage in oak-hickory stands in northeastern Oklahoma (Thompson and others 1991). Grasses and legumes increased after fire reduced the canopy in oak-pine stands in Oklahoma and Arkansas (Masters and others 1993). Understory burning at 1- and 2-year intervals favored herbaceous cover, while understory burning at 3- and 4-year intervals favored a mixture of herbs and shrubs.

Fire and Nutritional Quality—Based on a review of 16 studies of fire in Southern forests, Stransky and Harlow (1981) propose several generalizations about the effects of fire on plant nutrition. Burning typically increases the crude protein and phosphorus content of grasses, forbs, and browse the first postfire year. Increases in nutritive quality are greatest at the beginning of the growing season and decline rapidly, so protein and phosphorus levels are usually similar in burned and unburned areas by winter. Calcium content of plants after burning is highly variable among studies. The palatability of forage generally improves after fire, at least until growth stops and lignin content increases. Seldom are effects of burning on

Figure 22—Biennial prescribed burn plots, St. Marks National Wildlife Refuge, Florida. Area in foreground is burned every other August and dominated by runner oak, a mast-producing species. Area in background is burned every other April and shows wiregrass flowering. This is a species favored by seed-eating animals. Photo courtesy of Dale Wade.

the nutritional content of plants detected after the first growing season (Christensen 1977; Stransky and Harlow 1981; Thill and others 1987). However, high-intensity fires in the Florida Keys, oak communities, and pine-oak communities in the Southeast have extended fire's positive effects on plant nutrition for at least 1 year beyond the first growing season (Carlson and others 1993; DeWitt and Derby 1955; Thackston and others 1982). Exceptions to the pattern of nutrient increases after fire include reports from Florida sandridge habitat (Abrahamson and Abrahamson 1989) and eastern Texas pine-hardwood (O'Halloran and others 1987). These studies report no substantial increase in plant or fungus nutrient levels after fire.

Fire and the Quantity of Seeds and Mast—Seed and mast production generally increases after fire in Southern forests. According to Harlow and Van Lear (1989), seed production by legumes, grasses and spurges is significantly greater on annually burned areas than on areas burned less frequently. Production of berries, drupes, and pomes peaks 2 to 4 years after burning for most of 20 species of shrubs and small trees. A stand-replacement fire in pine and hardwood in the mountains of Virginia greatly increased the production of blueberries the second growing season after burning. Production declined by year 5 but remained higher than that on unburned plots (Coggins and Engle 1971). Season and frequency of burning influence berry production. Waldrop and others (1987) found that annual and biennial summer fires in loblolly pine forests on the Coastal Plain of South Carolina reduced the numbers of blueberry plants after 30 years of burning, whereas winter burning did not. According to a review by Robbins and Myers (1992), frequent growing season burns reduce mast-producing species except runner oak and some blueberry species.

Hard mast in the Southeast is used by a variety of birds and mammals, including northern bobwhite, wild turkey, sapsuckers, squirrels, black bear, and white-tailed deer. The current lack of frequent fires (both understory and stand-replacement) in the southern Appalachians is thought to be responsible for the replacement of oaks by other species (Van Lear 1991; Van Lear and Watt 1993). Small oaks resprout after fires, and large oaks have fire-resistant bark that enables these large trees to survive fire better than their competitors if frequently underburned. However, when fires are excluded for long periods, competing species such as tuliptree also develop fire-resistant bark. These competitors can survive fire, but they have much less potential than oaks for producing hard mast. Thirty years of prescribed burning in the Coastal Plain of South Carolina had no effect on the number of mast-producing hardwoods more than 5 inches

(12.5 cm) dbh (oaks, hickories, blackgum and others) (Waldrop and Lloyd 1991). Annual summer fires nearly eliminated small (less than 1 inch, 2.5 cm dbh) hardwoods, but all other burning treatments produced increases in the mast-producing species (Waldrop and others 1987).

Recommendations for use of prescribed fire often focus on specific wildlife-related objectives. Johnson and Landers (1978) recommend understory fire at 3-year intervals to optimize fruit production in open slash pines in Georgia, with some use of longer intervals to promote mast-producing species. The animals favored by this practice include white-tailed deer, common gray fox, northern bobwhite, wild turkey, raccoon, and songbirds. Hamilton (1981) recommends understory burning pine-hardwood habitats in winter at 5- to 10-year intervals to provide ample berries and mast for black bears. Harlow and Van Lear (1989) suggest that annual burning may not be desirable for the majority of wildlife species because mast production for most shrubs and small trees peaks 2 to 6 years after burning. However, annual burning would benefit northern bobwhite, mourning doves, some songbirds, and rodents.

Fire and Availability of Invertebrates—Fire effects on invertebrates relate not only to preburn conditions and fire severity but also to the life cycles and population patterns of the specific invertebrates studied. In the sandhills of the Florida panhandle, longleaf pine stands with dense turkey oak and sand live oak in the understory were underburned during the growing season. Arthropod density and biomass increased significantly, especially populations of grasshoppers, which constituted more than 90 percent of the arthropod biomass (Provencher and others 1998). Such increases are likely to benefit the northern bobwhite, a bird that feeds in the ground and herb layer, and hawking birds such as the loggerhead shrike and kestrels. Dunaway (1976) found that annual understory burning in longleaf pine did not increase the number of insects available to foraging birds, but may have provided open conditions conducive to successful hunting by northern bobwhite chicks. In central Florida sandhills, increased frequency of understory burning was positively correlated with the colony density of southern harvester ants (McCoy and Kaiser 1990). Understory fire in loblolly pine-shortleaf pine forest in east-central Mississippi increased invertebrates available to northern bobwhite and turkeys for up to 3 years (Hurst 1971, 1978).

While some insects are attracted to fire and increase rapidly in burns, fire reduces others. For example, fire is used in the southern Appalachian Mountains to control insects that prey on oak seedlings and mast (Van Lear and Watt 1993).

54

USDA Forest Service Gen. Tech. Rep. RMRS-GTR-42-vol. 1. 2000

Prairie Grassland

Fire and the Quantity of Forage—In presettlement times, frequent fires in grasslands kept tree cover in check. Studies in many regions describe the invasion of prairie by trees in the absence of fire (Gruell 1979; Reichman 1987; Sieg and Severson 1996). Where prairie fires eliminate trees, fires increase the amount of forage available to fauna simply by increasing the area covered by prairie. In addition, grassland fire can cause early green-up of warm-season grasses, improved seed germination, and greater production of grasses and forbs (Hulbert 1986, 1988; Svejcar 1990). Dramatic increases in yield during the first postfire year have been reported for dominant prairie grasses including prairie dropseed, big bluestem, western wheatgrass, bluebunch wheatgrass, and Indiangrass (Bushey 1987; Dix and Butler 1954; Hulbert 1988; Svejcar 1990). Many studies that report increased yield also describe some circumstances under which yield is reduced. In general, fires followed by drought and fires in areas with less than 11 inches (300 mm) of summer rainfall may cause decreased forage production (Kucera 1981).

Fire and Nutritional Quality—Fire often increases the percentage of protein and minerals in prairie grasses and shrubs, although effects vary with season of burning (Daubenmire 1968). Forage quality in mountain shrub and grassland communities is enhanced by increased availability of mineral nitrogen (Hobbs and Schimel 1984). The effects of fire on grassland nutrients interact with the effects of grazing. Grazed patches in a tallgrass prairie contained less biomass than ungrazed patches and therefore lost less nitrogen to volatilization by fire (Hobbs and others 1991). The differences were substantial enough that grazing may control whether burning causes net increases or decreases in nitrogen on a site. Grazing also increases heterogeneity in grasslands, contributing to patchy fuels and thus variation in fire behavior and severity.

Fire and Availability of Invertebrates—Reed (1997) reviewed studies of fire effects on prairie arthropod communities. She found that fire modified these communities, and the communities continued to change with time after fire. Prairies with fires initiated in different years and different seasons are likely to promote species richness. Fire in oak savannas, studied over a 30-year period, did not significantly alter arthropod diversity (Siemann and others 1997). Fires in Texas grasslands did not significantly alter arthropod abundance and availability to foraging birds (Koerth and others 1986). The density of arachnids and insect orders on Texas grasslands, however, differed significantly between burned and unburned areas at various times of year (table 5).

Beetle abundance declines immediately after fire in prairies but may return to prefire levels within a month (Rice 1932). On a tallgrass prairie in Kansas, arthropod biomass was greater on annually burned than unburned plots; cicada nymphs were more abundant on burned than unburned plots (Seastedt and

Table 5—Effects of fire on invertebrates in a Texas grassland after a January fire (Koerth and others 1986). Groups listed were significantly ($p < 0.05$) more (or less) abundant on burned than unburned areas, as indicated by density (number/m^2).

Month	First year after fire		Second year after fire	
	More abundant on burn	Less abundant on burn	More abundant on burn	Less abundant on burn
April			Orthoptera	Hemiptera
May		Hemiptera Homoptera Coleoptera		
June	Diptera	Arachnidae		Homoptera
July	Hymenoptera	Hemiptera Arachnidae		Homoptera
August				Coleoptera Arachnidae
September	Hymenoptera	Coleoptera Arachnidae		
October	Hymenoptera	Hemiptera Arachnidae	Orthoptera Arachnidae	

USDA Forest Service Gen. Tech. Rep. RMRS-GTR-42-vol. 1. 2000

55

others 1986). Cicadas respond positively to increased root productivity on burned sites, but they are relatively immobile so their feeding is unlikely to contribute to decline of their host plants.

Western Forests _____

Rocky Mountain Forest

Fire and the Quantity of Forage—Most studies of fire and wildlife foods in Western forests focus on ungulates. This research generally indicates that burning produces positive results for elk and mule deer. During the first 5 to 10 years following stand-replacing fire, grass and forb biomass generally increases. Grass and forb biomass decreased the first growing season after fire in aspen stands in Wyoming but increased the second and third growing seasons to above preburn levels (Bartos and Mueggler 1981). On "heavily burned" sites, grass recovered more slowly than forbs. Forage increased three-fold after both understory and stand-replacement fire in a ponderosa pine forest in Arizona (Oswald and Covington 1983). The increase persisted 9 years in underburned stands, but grazing, perhaps combined with severe fire effects, reduced forage after 2 years in areas burned by stand-replacing fire. Climax bunchgrass stands have been recommended for bighorn sheep winter range, but bighorn sheep in western Montana preferred seral forest with elk sedge and pinegrass openings (Riggs and Peek 1980).

Although total biomass of grasses and forbs often increases following fire, the quantity of useable forage may actually be less on burned areas if species composition shifts to domination by relatively unpalatable species. Prescribed understory burning failed to improve forage in some Southwestern ponderosa pine stands because, although herbage increased dramatically, flannel mullein, an unpalatable species, dominated the understory after fire (Ffolliott and Guertin 1990).

Burning brush fields in northern Idaho greatly increased the browse available to wintering elk the following year (Leege and Hickey 1971). In British Columbia, elk wintered primarily in postfire grass and shrub communities, except during severe weather when conifer stands were used (Peck and Peek 1991). In Idaho, mule deer foraged primarily in burned habitats in winter, while white-tailed deer foraged primarily in unburned habitats (Keay and Peek 1980). An intense prescribed fire in Douglas-fir in Idaho improved forage for mule deer and elk. The benefits were expected to last more than 20 years (Lyon 1971). Positive effects of fire on grazing and browse productivity generally last less than 30 years (Oswald and Covington 1983; Pearson and others 1972).

Mixed-severity and stand-replacement fires often increase berry-producing shrubs and their productivity 20 to 60 years after fire. These changes benefit birds, small mammals, and bear. Increased production of forb foliage and tuberous roots after the 1988 Yellowstone fires benefited grizzly bears (Blanchard and Knight 1996). A mathematical model predicts increased wintering populations of elk and bison in Yellowstone for 20 to 30 years postfire (Boyce and Merrill 1991). Large, intense burns may be necessary for long-term maintenance of natural forest succession patterns of some forest types and for habitat diversity in others (Finch and others 1997). While fires top-kill huckleberry plants and kill many whitebark pines, two species that provide important forage for grizzly bears, they also rejuvenate decadent huckleberry stands and prevent subalpine fir from replacing whitebark pine in many high elevation forests (Agee 1993).

Fire and Nutritional Quality—Fires usually increase some nutrients in Rocky Mountain forests and the pine forests of Arizona and New Mexico for 1 to 3 years (Severson and Medina 1983). Stand-replacement fall burning in Wyoming aspen stands increased crude protein and phosphorus of forage during the first summer after treatment (DeByle and others 1989). In vitro dry matter digestibility was also higher in burned areas, and calcium content was lower. By late summer, only crude protein levels were different and, in the second postfire year, forage quality was similar on burned and unburned areas. Burning improves the nutritional qualities of forage plants in ponderosa pine forest for one to three growing seasons (Meneely and Schemnitz 1981; Pearson and others 1972; Rowland and others 1983). In western larch/Douglas-fir stands in Montana that had been burned with understory fire 3 years previously, nutrient content of plants was compared with samples from stands not burned for 70 years (Stark and Steele 1977). Sodium levels were higher for several species in stands where at least half of the duff was consumed by fire. Iron concentration was significantly greater in some species on burned than unburned sites, and calcium and phosphorus were significantly lower. The plant species tested showed no significant differences in nitrogen, magnesium, or copper between burned and unburned sites. Scouler's willow in underburned ponderosa pine/Douglas-fir forests in Montana contained higher concentrations of phosphorus and crude protein, and lower lignin concentration, than willows in unburned stands (Bedunah and others 1995).

Some research reports no significant changes in nutrient levels after fire. Seip and Bunnell (1985) found no differences in the nutritive quality of forage on frequently burned alpine range and unburned

56

range used by Dall's sheep in British Columbia. The authors thought that sheep on burned range were in better physical condition than those on unburned range because of the quantity of forage rather than its nutritive quality. Stand-replacing prescribed fire in Idaho aspen forests in September produced little change in the nutritive content of forage the first, second, and fourth growing seasons after burning (Canon and others 1987). However, elk preferred to forage in the burned areas, possibly because preferred species were consistently available and foraging was more efficient.

Sierra Forest

Fire and the Quantity of Forage and Seed— Wildlife forage species in Sierra Nevada forests include many plants that dominate in chaparral to the west and more mesic forests to the north. Deerbrush and greenleaf manzanita are chaparral species but are also important components of the understory of Sierra forests. Forage of deerbrush and other *Ceanothus* species, which is high quality food for ungulates (Sampson and Jesperson 1963; Stubbendieck and others 1992), is abundant after fire because it reproduces from seed that is scarified by burning (Burcham 1974). Early spring burning in the Sierra Nevada increases palatability of foliage for wildlife (Kauffman and Martin 1985). Thimbleberry is an understory species characteristic of mesic Sierra forests; it generally increases after fire (Hamilton and Yearsley 1988).

Pacific Coast Maritime Forest

Fire and the Quantity of Forage—Salmonberry, an important understory species in Pacific Coast forests, is used by numerous wildlife species. Deer, elk, mountain goats, and moose browse on its buds and twigs; songbirds, gallinaceous birds, bears, and coyote feed on its fruit. Salmonberry sprouts prolifically and grows rapidly in the first years after fire, although severe fire may reduce sprouting (Tappeiner and others 1988; Zasada and others 1989).

Western Woodlands, Shrublands, and Grasslands _____

Pinyon-Juniper

Fire and the Quantity of Forage—Severson and Medina (1983) and Severson and Rinne (1990) review the effects of fire on forage production and wildlife habitat in the Southwest. While they demonstrate the important role of fire in improving Southwestern vegetation types for wildlife, they emphasize the need for a balance between burned and unburned areas. Fire intensity varies greatly in pinyon-juniper woodlands, and the early successional effects of fires are difficult to predict (Severson and Rinne 1990). Often, fire may not have much effect unless combined with other treatments (Wittie and McDaniel 1990). When conditions are favorable for stand-replacing fire, burning kills most of the pinyon-juniper overstory and increases diversity in the plant community, with some effects lasting up to 115 years after fire (McCulloch 1969; Severson and Medina 1983; Severson and Rinne 1990; Stager and Klebenow 1987). Shortly after fire, burns are usually dominated by forbs, with grasses becoming abundant a few years later. In an Ashe's juniper community burned during a moist winter and spring, grasses recovered quickly and soil erosion was minimal (Wink and Wright 1973). Similar treatments during a dry winter and spring, however, reduced herbaceous yields and increased erosion.

Chaparral and Western Oak Woodlands

Fire and the Quantity of Forage—Intense fires in chaparral result in a flush of herbaceous plants and shrubs for 1 to 5 years (Biswell 1974; Christensen and Muller 1975; Klinger and others 1989; Taber and Dasmann 1958). In Gambel oak rangeland in Colorado, fire did not significantly change the biomass of forbs and shrubs 2, 5, and 10 years after fall mixed-severity fire, but grass biomass was greater on burned than unburned sites during postfire year 10 (Kufeld 1983).

Fire and Nutritional Quality—Most studies of postfire nutrients in Western ecosystems report some changes, but the plant species and the nutrients affected vary. Stand-replacing fires in chaparral increased the protein content of leaves for one to two growing seasons and the phosphorus content for up to 6 years (Rundel and Parsons 1980; Taber and Dasmann 1958). Two growing seasons after fall mixed-severity burns in Gambel oak rangeland in Colorado, zinc and copper levels were higher in plants on burned than unburned sites. However, no differences were found in the protein, lignin, calcium, or phosphorus content of forbs, grasses or shrubs growing on burned and unburned areas (Kufeld 1983).

Where postfire nutrient changes vary among the plant species available to fauna, animals may select the more nutritious foods. September prescribed burns in mountain shrub and grassland habitats in Colorado increased the level of protein and in vitro digestible organic matter in winter diets of bighorn sheep and mule deer (Hobbs and Spowart 1984). Burning had no detectable effect on spring diets. The effects of burning on crude protein in the diet persisted for 2 years in both communities. The effect on digestible matter was present only in the mountain shrub habitat the second year. The increase in the nutritional quality of diets

USDA Forest Service Gen. Tech. Rep. RMRS-GTR-42-vol. 1. 2000

57

was greater than the apparent increase in the quality of browse and forage, indicating that sheep and deer foraged selectively for the plants that were more nutritious.

Fire and the Quantity of Seeds and Mast—The acorns produced by Western oak woodlands are used by birds, small mammals, and ungulates. Oaks that have been severely damaged by fire may produce "massive" seed crops (Rouse 1986).

Sagebrush and Sagebrush Grasslands

Fire and the Quantity of Forage—Some studies report no increases in grass and sagebrush productivity due to fire but do report other changes favorable to ungulates. Burning big sagebrush-bluebunch wheatgrass winter range in Wyoming decreased sagebrush for the 4 years of study but did not increase wheatgrass. Annual forbs were more abundant on the burned area only the second year after burning. Nonetheless, bighorn sheep and possibly mule deer made greater use of the burned areas than the unburned areas (Peek and others 1979). Prescribed burning reduced plant litter that inhibited grazing by elk on a Montana fescue-wheatgrass winter range. Fire did not significantly change the forb, shrub, and grass standing crops, however, except that rough fescue, the preferred winter forage, was reduced the first year after burning (Jourdonnais and Bedunah 1990).

Fire and Availability of Invertebrates—Stand-replacing and mixed-severity fire in big sagebrush in Oregon did not affect populations of darkling beetles or June beetles (Pyle and Crawford 1996). Fire did not appreciably alter their food and cover on the ground surface.

Deserts

Fire and the Quantity of Forage—Fire reduces most shrubs in the Great Basin Desert for at least a few years (Humphrey 1974). In the first year after fire, perennial grasses and forbs have reduced vigor and annuals are abundant. By the third year, total herbage often reaches a maximum, exceeding production on unburned sites, and grasses and herbs flower profusely. The dominant grasses are thickspike wheatgrass, plains reedgrass, and bluebunch wheatgrass; other grasses, including bluegrass species and Idaho fescue, do not recover to preburn production until the second decade after fire.

Fire in the Mojave Desert is likely only after a season of heavy production by annual plants. The moisture levels of woody and perennial plants determines the level of mortality. If conditions are excessively dry, damage is severe.

In the Sonoran and Chihuahuan Deserts, fire is uncommon because of the widely spaced, open-branched vegetation. In wet years, fires occur in grasslands and their interface with desert, killing woody plants, such as velvet mesquite, and expanding the grassland. Fires that burn off the spines from cacti (cholla, pricklypear, and barrel cactus) make the plants available as forage for livestock and rabbits. Fires at the grassland-woodland ecotone may remove woody vegetation without increasing ground cover (Bock and Bock 1990). In desert grasslands, fire is likely to reduce yield for 1 to 2 years, with productivity recovering to preburn levels by the third year (Jameson 1962; Wright 1980). Where black grama is dominant, fire effects vary. Productivity may be reduced for 10 years or longer (Wright 1980). Tobosagrass production increased two to threefold after early spring burns followed by rain, but burning in a dry spring reduced yield (Wright 1973).

Subtropical Ecosystems _____

Florida Wetlands

Fire and the Quantity of Forage—In Florida wetlands, fires increase open aquatic areas and reduce the encroachment of pine hammocks, thus altering the balance between terrestrial and aquatic habitat. Burning opens up cattail stands by removing years of accumulated litter. Fire eliminates litter in sawgrass stands and reduces plant height for a year or two. To maintain fruit production for white-tailed deer, Fults (1991) recommends burning saw-palmetto understories every 3 to 5 years.

58

USDA Forest Service Gen. Tech. Rep. RMRS-GTR-42-vol. 1. 2000

L. Jack Lyon
Jane Kapler Smith

Chapter 8: Management and Research Implications

Management Implications _____

Only a few places in North America, or the world, exist where fire has not shaped the vegetation or influenced the faunal community. In many areas of North America, managers have successfully prevented or limited the occurrence of this natural process for nearly 100 years, and that century of fire exclusion has probably caused many changes in habitat and wildlife populations of which we are not even aware. It is likely that some faunal populations and communities present in today's landscapes could not have developed under pre-1900 fire regimes. Many researchers and managers agree, however, that the success of fire exclusion cannot continue (Fiedler and others 1998; Fule and Covington 1995) and, indeed, is already beginning to fail (Barbouletos and others 1998; Wicklow-Howard 1989; Williams and others 1998). Fire is most likely to increase in wildlands in the future. This likelihood carries with it two broad implications for the relationships between fire and fauna.

One: Alternatives in Managing Fire

Managers are increasingly likely to have to choose among:

- Massive fire suppression (with increasing hazards and increasing costs).
- Uncontrolled, possibly uncontrollable fires.
- A combination of prescribed fires and wildland fires used to achieve resource objectives.

The implications of these choices for animal communities in North American wildlands are significant. Most North American fauna communities have developed under pressure from repeated fires of specific severities and frequencies. Alteration of that pressure for the past 100 to 500 years has changed the abundance and geographic distribution of many kinds of habitat and the animals that depend on it.

Even more important than changes in past centuries, however, is the likelihood that fires in the immediate future will deviate substantially from what might be considered normal or natural in many areas of North America. While restoration of presettlement fire regimes may be desirable for habitat protection, this may be impossible in many areas because of fuel accumulation, structural change due to fire exclusion, and climate change (see discussion of this topic in "Effects of Fire on Flora" in the Rainbow Series). Even if habitat restoration is successful, animal populations may be slow to colonize treated areas, so

USDA Forest Service Gen. Tech. Rep. RMRS-GTR-42-vol. 1. 2000

59

perpetuation of existing habitat is a more reliable management strategy than restoration of degraded habitat. Managers attempting to restore habitat by emulating presettlement fire regimes will not only encounter increased fuel loads and increased continuity of fuels, but also resistance from the public because of the immediate increased risks to human life, health, property, and welfare. The altered vegetation may need to be burned under conditions that would not normally incur extensive fire spread. For many fauna species, this practice can produce site and landscape conditions completely outside the range of those under which the species evolved. Because spatial and temporal variation are important aspects of presettlement fire regimes, management plans should address these features explicitly whenever possible (Lertzman and others 1998).

Considering the many variables and unknowns that impinge upon management choices in regard to fire, careful consideration of the science and monitoring of treatment results is important. As Rieman and others (1997) comment regarding fire effects on aquatic fauna, "There is undoubtedly a point where the risk of fire outweighs the risk of our management, but that point needs to be discovered through careful evaluation and scientific study not through the opposing powers of emotional or political rhetoric."

Two: Integrating Management Objectives

Objectives of prescribed fires and use of wildland fires for resource benefits must be clearly stated and integrated with overall land management objectives, addressing the potential for interaction among disturbances such as grazing, flood, windthrow, predation, and insect and fungal infestation. In the past 10,000 years, fire has never operated in isolation from other disturbances, nor has fire usually occurred independent of human influence (Kay 1998; Pyne 1982). During thousands of years prior to settlement of North America by European Americans, Native Americans influenced both fire regimes and animal populations. In fact, populations of large ungulates may have been limited by Native American predation rather than food (Kay 1998). As Kay (1995) states, "Setting aside an area as wilderness or a National Park today, and then managing it by letting nature take its course will not preserve some remnant of the past but instead create conditions that have not existed for the last 10,000 yr." As managers face ubiquitous needs for addressing fire in land management, and as they encounter increasing difficulty in managing habitat in conditions near those under which faunal species evolved, we believe it is of paramount importance to have clear objectives for use of prescribed fire, wildland fire for resource benefits, and fire suppression, based on understanding of past disturbance patterns

and human influence. It is important to avoid, if possible, major deviations into ecological conditions outside the range of variability that occurred in the millennium prior to 1900.

When fire suppression and use are not integrated with overall management programs, the potential for unanticipated problems and failure increases. Management for aspen restoration and bighorn sheep range improvement provide two examples. If aspen is treated by fire to regenerate the stand but then repeatedly browsed by wildlife, it often deteriorates more rapidly than without treatment (Bartos 1998; Basile 1979). The choice of treatment and the size and distribution of treated sites must in this case be integrated with knowledge of wildlife use patterns and wildlife management. Prescribed fire can negatively affect bighorn sheep habitat when range condition is already poor, when the burn leaves inadequate forage for the winter, and when other species, especially elk, are attracted to the burned habitat (Peek and others 1985). Again, fire management needs to be integrated with wildlife information and management.

Understanding of fire history, potential fire behavior, and differing needs of multiple species must be integrated in planning for prescribed fire. For example, since many small mammals use tunnels under forest litter and in or near large pieces of dead wood as refugia (Ford and others 1999), managers can influence the impact of fire on small mammals by including moisture levels of these fuels in plans for fire use. Salvaged logged sites in stand-replacement burns in the Northern Rocky Mountains provide nesting opportunities for some cavity nesters (northern flicker, hairy woodpecker, and mountain bluebird). Other bird species (black-backed woodpecker, northern three-toed woodpecker, and brown creeper) occur almost exclusively in burned, unlogged patches (Hejl and McFadzen 1998). If salvage logging is considered after a wildland fire, the needs of the specific bird community in the area must be considered.

Because funding and other resources for management will always be limited, it is important to use objectives to shape clear priorities for fire suppression and fire use. Is it more important to use limited resources on small areas that will benefit small, but perhaps irreplaceable, populations of animals? Or is it more important to restore large areas and address the challenges of landscape-level planning? Only carefully thought-out objectives can guide such choices well.

Needs for Further Understanding ___

Research questions regarding fire effects on fauna fall into two categories: (1) those regarding fauna-habitat relationships and (2) those regarding presettlement fire regimes.

60

USDA Forest Service Gen. Tech. Rep. RMRS-GTR-42-vol. 1. 2000

Fauna-Habitat Relationships

Information involving relationships between fire and animals is needed for all classes of fauna. Most of the information currently available focuses on vertebrates, particularly mammals and birds. Studies of landscape and community ecology are virtually limited to birds. Furthermore, most studies are limited to population descriptors, while measurement of productivity may be essential for understanding fire effects and predicting effects of management options. Given the relative lack of information about fire effects on herpetofauna and insects, studies in those areas may be especially important (Pickering 1997; Russell and others 1999). Future research should address microsite conditions, patchiness within burns, and seasonality of fire effects for specific ecosystems. Likewise, information about fire effects on aquatic fauna is sparse, much of it originating from only a few ecosystems (for example, see Bozek and Young 1994; Mihuc and others 1996; Minshall and others 1989; Rieman and others 1997). More information is needed regarding long-term effects, landscape effects, and effects of postfire succession on aquatic fauna. (See also discussion of this topic in "Effects of Fire on Soil and Water" in the Rainbow Series.)

The need to fill information gaps will increase as stands and landscapes continue to diverge from pre-settlement patterns and as managers increasingly use fire for vegetation management. To improve long-term management for sustaining ecosystems, information is needed about the effects of fire on many kinds of fauna, at different seasons and under different conditions, and over many decades. Information on the interactions of burning season with life cycles of animal species, especially insects and herpetofauna, is also important.

Site-Level Research Questions—At the site level, managers need detailed information on the use of fire to manage the structure of vegetation, especially in shrublands and forest understories. Objectives for this kind of management include maintaining nesting habitat for birds, ensuring habitat features needed for reproduction by herpetofauna and insects, providing cover for small mammals, and enhancing local community diversity.

Also at the site level, managers need better designed, more comprehensive studies of fire impacts on quantity and quality of forage for wildlife. A truly vast literature addresses this subject, but much of it is hard to apply because the investigators did not control for factors other than burning and did not describe fire severity or burning conditions in detail. Land managers in many localities currently use limited amounts of prescribed fire to enhance wildlife habitat, but more widespread use of fire in habitat management will require more comprehensive knowledge than is currently available.

Landscape-Level Research Questions—At the landscape level, we lack almost any knowledge of the combination of mosaics and patterns best suited to specific populations, and we have little understanding of how to maintain the total landscape for regional biodiversity. While habitat corridors are important for sustaining some wildlife species (Beier and Noss 1998; Oliver and others 1998), what are the implications of fire and succession in corridors and the locations that provide access to them? Some research of this kind is under way, but limitations of time and money will virtually assure that computer models rather than landscape-level experiments will provide the greatest progress (Schmoldt and others 1999).

Wildlife researchers often face a dilemma regarding research priorities: Should we invest time and resources in learning more about faunal habitat, or should we learn more about the species themselves? The answer depends on the ecosystem under study. Schultz and Crone (1998) developed a model for habitat change in the prairie habitat of the Fender's blue butterfly, a candidate for listing on the U.S. Endangered Species list. They report that lack of knowledge about postfire habitat change limited the certainty of the model's predictions more than lack of knowledge about the butterfly itself. In contrast, both Wright (1996) and Telfer (1993) state that information about the fauna species investigated (birds in both studies), especially nesting success, currently limits our ability to understand the effects of potential management choices, including those regarding fire.

Presettlement Fire Regimes

Important knowledge gaps remain about the distribution and structure of vegetation in presettlement times. Without this information, managers cannot decide what proportion of forest land should be in various age classes, structural classes, and cover types to maintain biodiversity. Furthermore, managers need methods for integrating current agricultural and infrastructural elements in the landscape with remaining wildlands at large scales, approximating the original fire-shaped mosaic and structure for an area as well as possible. With this information, wildlands can be used to the best advantage to maintain regional biodiversity, increase numbers of particular wildlife species, and achieve other environmental goals.

Human Dimension _____

Finally, researchers and managers need to collaborate in assessing the comparative merits and drawbacks of various kinds of fire for natural resource

USDA Forest Service Gen. Tech. Rep. RMRS-GTR-42-vol. 1. 2000

61

objectives across the landscape. What ecological and social risks occur with prescribed fires, wildland fires managed for resource objectives, and fire suppression? How can these risks be reduced? It is impossible to know all the consequences of intervening in an ecosystem, whether the intervention is active (prescribed fire, for example), or passive (such as fire exclusion or landscape fragmentation). Monitoring and comparison of monitoring results with predictions are essential. Communication among researchers, managers, and the public is also essential. Science cannot be used until it is shared with and understood by managers, whose job is to apply the results, and a substantial proportion of the public, who add the perspective of their values and experience. Policy, according to Pyne (1982), "has to be based on broad cultural perceptions and political paradigms, not solely on ecological or economic investigations; scientific research is only one component among many that contribute to it."

62

USDA Forest Service Gen. Tech. Rep. RMRS-GTR-42-vol. 1. 2000

References

Abrahamson, Warren G.; Abrahamson, Christy R. 1989. Nutritional quality of animal dispersed fruits in Florida sandridge habitats. Bulletin of the Torrey Botanical Club. 116(3): 215-228.

Agee, James K. 1993. Fire ecology of Pacific Northwest forests. Washington, DC: Island Press. 493 p.

Agee, James K. 1998. The landscape ecology of western forest fire regimes. Northwest Science. 72(special issue): 24-34.

Ahlgren, C. E. 1974. Fire and ecosystems—introduction. In: Kozlowski, T. T.; Ahlgren, C. E., eds. Fire and ecosystems. New York, NY: Academic Press: 1-5.

Allaby, Michael, ed. 1992. The concise Oxford dictionary of botany. New York, NY: Oxford University Press. 442 p.

Anderson, H. E. 1983. Predicting wind-driven wildland fire size and shape. Res. Pap. INT-305. Ogden, UT: U.S. Department of Agriculture, Forest Service, Intermountain Research Station. 26 p.

Apfelbaum, Steven; Haney, Alan. 1981. Bird populations before and after wildfire in a Great Lakes pine forest. Condor. 83: 347-354.

Arno, Stephen F. 1976. The historical role of fire on the Bitterroot National Forest. Res. Pap. INT-42. Ogden, UT: U.S. Department of Agriculture, Forest Service, Intermountain Forest and Range Experiment Station. 28 p.

Ashcraft, G. C. 1979. Effects of fire on deer in chaparral. Cal-Neva Wildlife Transactions. 1979: 177-189.

Auclair, A. N. D. 1983. The role of fire in lichen-dominated tundra and forest-tundra. In: Wein, Ross W.; MacLean, David A., eds. The role of fire in northern circumpolar ecosystems. New York, NY: John Wiley and Sons: 235-255.

Barbouletos, Catherine S.; Morelan, Lynette Z.; Carroll, Franklin O. 1998. We will not wait: why prescribed fire must be implemented on the Boise National Forest. In: Pruden, Teresa L.; Brennan, Leonard A., eds. Fire in ecosystem management: shifting the paradigm from suppression to prescription: Proceedings, 20th Tall Timbers fire ecology conference; 1996 May 7-10; Boise, ID. Tallahassee, FL: Tall Timbers Research Station: 27-30.

Barney, Milo A.; Frischknecht, Neil C. 1974. Vegetation changes following fire in the pinyon-juniper type of west-central Utah. Journal of Range Management. 27(2): 91-96.

Barron, M. G. 1992. Effect of cool and hot prescribed burning on breeding songbird populations in the Alabama piedmont. Auburn, AL: Auburn University. Thesis. 39 p.

Bartos, Dale. 1998. Aspen, fire and wildlife. In: Fire and wildlife in the Pacific Northwest—research, policy, and management; 1998 April 6-8; Spokane, WA: 44-48.

Bartos, Dale L.; Mueggler, W. F. 1981. Early succession in aspen communities following fire in western Wyoming. Journal of Range Management. 34(4): 315-318.

Basile, Joseph V. 1979. Elk-aspen relationships on a prescribed burn. Res. Note INT-271. Ogden, UT: U.S. Department of Agriculture, Forest Service. 7 p.

Bedunah, Donald J.; Willard, E. Earl; Marcum, C. Les. 1995. Response of willow and bitterbrush to shelterwood cutting and underburning treatments in a ponderosa pine forest. Final report: Research Joint Venture Agreement No. INT-92684-RJVA. On file at: U.S. Department of Agriculture, Forest Service, Rocky Mountain Research Station, Fire Sciences Laboratory, Missoula, MT. 36 p.

Beier, Paul; Noss, Reed F. 1998. Do habitat corridors provide connectivity? Conservation Biology. 12(6): 1241-1252.

Belsky, A. Joy. 1996. Western juniper expansion: Is it a threat to arid northwestern ecosystems? Journal of Range Management. 49: 53-59.

Bendell, J. F. 1974. Effects of fire on birds and mammals. In: Kozlowski, T. T.; Ahlgren, C. E., eds. Fire and ecosystems. New York, NY: Academic Press: 73-138.

Benson, Lee A.; Braun, Clait E.; Leininger, Wayne C. 1991. Sage grouse response to burning in the big sagebrush type. In: Comer, Robert D.; Davis, Peter R.; Foster, Susan Q.; Grant, C. Val; Rush, Sandra; Thorne, Oakleigh, II; Todd, Jeffrey, eds. Issues and technology in the management of impacted wildlife: Proceedings of a national symposium; 1991 April 8-10; Snowmass Resort, CO. Boulder, CO: Thorne Ecological Institute: 97-104.

Bentz, Jerry A.; Woodard, Paul M. 1988. Vegetation characteristics and bighorn sheep use on burned and unburned areas in Alberta. Wildlife Society Bulletin. 16(2): 186-193.

Best, L. B. 1979. Effects of fire on a field sparrow population. American Midland Naturalist. 101(2): 434-442.

Bevis, Kenneth R.; King, Gina M.; Hanson, Eric E. 1997. Spotted owls and 1994 fires on the Yakama Indian Reservation. In: Greenlee, Jason M., ed. Proceedings, 1st conference on fire effects on rare and endangered species and habitats; 1995 November 13-16; Coeur d'Alene, ID. Fairfield, WA: International Association of Wildland Fire: 117-122.

Beyers, Jan L.; Wirtz, William O., II. 1997. Vegetative characteristics of coastal sage scrub sites used by California gnatcatchers: implications for management in a fire-prone ecosystem. In: Greenlee, Jason M., ed. Proceedings, 1st conference on fire effects on rare and endangered species and habitats; 1995 November 13-16; Coeur d'Alene, ID. Fairfield, WA: International Association of Wildland Fire: 81-89.

Bidwell, T. G. 1994. Effects of introduced plants on native wildlife in the Great Plains. Riparian area management: proceedings of the 46th annual meeting, forestry committee, Great Plains agricultural council; 1994 June 20-23; Manhattan, KS. Publication no. 149. Manhattan, KS: Great Plains Agricultural Council: 73-79.

Biswell, H. H. 1961. Manipulation of chamise brush for deer range improvement. California Fish and Game. 47(2): 125-144.

Biswell, H. H. 1963. Research in wildland fire ecology. In: Proceedings, 2nd annual Tall Timbers fire ecology conference; 1963 March 14-15; Tallahassee, FL. Tallahassee, FL: Tall Timbers Research Station: 63-97.

Biswell, H. H. 1974. Effects of fire on chaparral. In: Kozlowski, T. T.; Ahlgren, C. E., eds. Fire and ecosystems. New York, NY: Academic Press: 321-364.

Biswell, Harold H. 1989. Prescribed burning in California wildlands vegetation management. Berkeley, CA: University of California Press. 255 p.

Blanchard, Bonnie; Knight, Richard R. 1996. Effects of wildfire on grizzly bear movements and food habits. In: Greenlee, Jason M., ed. The ecological implications of fire in Greater Yellowstone: Proceedings, 2nd biennial conference on the Greater Yellowstone Ecosystem; 1993 September 19-21; Yellowstone National Park, WY. Fairfield, WA: International Association of Wildland Fire: 117-122.

Blankenship, Daniel J. 1982. Influence of prescribed burning on small mammals in Cuyamaca Rancho State Park, California. In: Conrad, C. Eugene; Oechel, Walter C., eds. Proceedings of the symposium on dynamics and management of Mediterranean-type ecosystems; 1981 Jun 22-26; San Diego, CA. Gen. Tech. Report PSW-58. Berkeley, CA: U.S. Department of Agriculture, Forest Service, Pacific Southwest Forest and Range Experiment Station: 587.

Bock, Carl E.; Bock, Jane H. 1978. Response of birds, small mammals, and vegetation to burning sacaton grasslands in southeastern Arizona. Journal of Range Management. 31(4): 296-300.

Bock, Carl E.; Bock, Jane H. 1983. Responses of birds and deer mice to prescribed burning in ponderosa pine. Journal of Wildlife Management. 47(3): 836-840.

Bock, Carl E.; Bock, Jane H. 1987. Avian habitat occupancy following fire in a Montana shrubsteppe. Prairie Naturalist. 19(3): 153-158.

Bock, Carl E.; Bock, Jane H. 1990. Effects of fire on wildlife in southwestern lowland habitats. In: Krammes, J. S., tech. coord. Effects of fire management of southwestern natural resources: Proceedings; 1988 Nov. 15-17; Tucson, AZ. Gen. Tech. Rep. RM-191. Fort Collins, CO: U.S. Department of Agriculture, Forest Service, Rocky Mountain Forest and Range Experiment Station: 50-64.

USDA Forest Service Gen. Tech. Rep. RMRS-GTR-42-vol. 1. 2000

63

Bock, Carl E.; Bock, Jane H. 1992. Response of birds to wildfire in native versus exotic Arizona grassland. Southwestern Naturalist. 37(1): 73-81.

Bock, Carl E.; Lynch, J. F. 1970. Breeding bird populations of burned and unburned conifer forest in the Sierra Nevada. Condor. 72: 182-189.

Bock, Carl E.; Raphael, M.; Bock, Jane H. 1978. Changing avian community structure during early post-fire succession in the Sierra Nevada. Wilson Bulletin. 90: 119-123.

Bone, Steven D.; Klukas, Richard W. 1990. Prescribed fire in Wind Cave National Park. In: Alexander, M. E.; Bisgrove, G. F., tech. coords. The art and science of fire management, proceedings of the First Interior West Fire Council Annual Meeting and Workshop; 24-27 October 1988; Kananaskis Village, AB. Information Report NOR-X-309. Edmonton, AB: Forestry Canada, Northwest Region, Northern Forestry Centre: 297-302.

Borror, Donald J.; White, Richard E. 1970. A field guide to insects, America north of Mexico. Boston, MA: Houghton Mifflin Company. 404 p.

Boyce, Mark S.; Merrill, Evelyn H. 1991. Effects of the 1988 fires on ungulates in Yellowstone National Park. In: High intensity fire in wildlands: management challenges and options: Proceedings, 17th fire ecology conference; 1989 May 18-21; Tallahassee, FL. Tallahassee, FL: Tall Timbers Research Station: 121-132.

Bozek, Michael A.; Young, Michael K. 1994. Fish mortality resulting from delayed effects of fire in the Greater Yellowstone Ecosystem. Great Basin Naturalist. 54(1): 91-95.

Bradley, A. F.; Noste, N. V.; Fischer, W. C. 1992. Fire ecology of forests and woodlands in Utah. Gen. Tech. Rep. INT-287. Ogden, UT: U.S. Department of Agriculture, Forest Service, Intermountain Research Station. 128 p.

Breininger, David R.; Larson, Vickie L.; Duncan, Brean W.; Smith, Rebecca B.; Oddy, Donna M.; Goodchild, Michael R. 1995. Landscape patterns of Florida scrub jay habitat use and demographic success. Conservation Biology. 9(6): 1442-1453.

Breininger, David R.; Larson, Vickie L.; Oddy, Donna M.; Smith, Rebecca B. [In press]. How does variation in fire history influence Florida scrub-jay demographic success? In: Greenlee, Jason, ed. 2nd conference, fire effects on rare and endangered species; 29 March-1 April 1998; Coeur d'Alene, ID. Fairfield, WA: International Association of Wildland Fire.

Breininger, David R.; Schmalzer, P. A. 1990. Effects of fire on plants and birds in a Florida oak/palmetto scrub community. American Midland Naturalist. 123: 64-74.

Breininger, David R.; Smith, Rebecca B. 1992. Relationships between fire and bird density in coastal scrub and slash pine flatwoods in Florida. American Midland Naturalist. 127: 233-240.

Brown, Arthur A.; Davis, Kenneth P. 1973. Forest fire control and use, 2nd edition. New York, NY: McGraw-Hill. 686 p.

Brown, James K.; DeByle, Norbert V. 1982. Developing prescribed burning prescriptions for aspen in the Intermountain West. Proceedings of the symposium: Fire—its field effects; 1982 October 19-21; Jackson, WY. Missoula, MT: Intermountain Fire Council; Pierre, SD: South Dakota Division of Forestry, Rocky Mountain Fire Council: 29-49.

Brown, James K.; Oberheu, Rick D.; Johnston, Cameron M. 1982. Handbook for inventorying surface fuels and biomass in the Interior West. Gen. Tech. Rep. INT-129. Ogden, UT: U.S. Department Agriculture, Forest Service, Intermountain Forest and Range Experiment Station. 48 p.

Brown, Timothy K.; Bright, Larry. 1997. Wildlife habitat preservation and enrichment during and after fires. In: Greenlee, Jason M., ed. Proceedings, 1st conference on fire effects on rare and endangered species and habitats; 1995 November 13-16; Coeur d'Alene, ID. Fairfield, WA: International Association of Wildland Fire: 65-68.

Buell, Murray F.; Cantlon, John E. 1953. Effects of prescribed burning on ground cover in the New Jersey pine region. Ecology. 34(3): 520-528.

Bull, Evelyn L.; Blumton, Arlene K. 1999. Effect of fuels reduction on American martens and their prey. Res. Note PNW-RN-539. Portland, OR: U.S. Department of Agriculture, Forest Service, Pacific Northwest Research Station. 9 p.

Bull, Evelyn L.; Torgersen, Torolf R.; Blumton, Arlene K.; McKenzie, Carol M.; Wyland, Dave S. 1995. Treatment of an old-growth stand and its effects on birds, ants, and large woody debris: a case study. Gen. Tech. Rep. PNW-GTR-353. Portland, OR: U.S. Department of Agriculture, Forest Service, Pacific Northwest Research Station. 12 p.

Bunnell, Fred L. 1995. Forest-dwelling vertebrate faunas and natural fire regimes in British Columbia: patterns and implications for conservation. Conservation Biology. 9(3): 636-644.

Burcham, L. T. 1974. Fire and chaparral before European settlement. In: Rosenthal, Murray, ed. Symposium on living with the chaparral: Proceedings; 1973 March 30-31; Riverside, CA. San Francisco, CA: The Sierra Club: 101-120.

Bushey, Charles L. 1987. Short-term vegetative response to prescribed burning in the sagebrush/grass ecosystem of the northern Great Basin; three years of postburn data from the demonstration of prescribed burning on selected Bureau of Land Management districts. Missoula, MT: Systems for Environmental Management. Final Report: Cooperative Agreement 22-C-4-INT-33. 77 p.

Cable, D. R. 1967. Fire effects on semidesert grasses and shrubs. Journal of Range Management. 20: 170-176.

Campbell, Bruce H.; Hinkes, Mike. 1983. Winter diets and habitat use of Alaska bison after wildfire. Wildlife Society Bulletin. 11(1): 16-21.

Canon, S. K.; Urness, P. J.; DeByle, N. V. 1987. Habitat selection, foraging behavior, and dietary nutrition of elk in burned aspen forest. Journal of Range Management. 40(5): 433-438.

Carlile, Lawrence D. 1997. Fire effects on threatened and endangered species and habitats of Fort Stewart Military Reservation, Georgia. In: Greenlee, Jason M., ed. Proceedings, 1st conference on fire effects on rare and endangered species and habitats; 1995 November 13-16; Coeur d'Alene, ID. Fairfield, WA: International Association of Wildland Fire: 227-231.

Carlson, Peter C.; Tanner, George W.; Wood, John M.; Humphrey, Stephen R. 1993. Fire in Key deer habitat improves browse, prevents succession and preserves endemic herbs. Journal of Wildlife Management. 57(4): 914-928.

Castrale, J. S. 1982. Effects of two sagebrush control methods on nongame birds. Journal of Wildlife Management. 46(4): 945-952.

Caton, Elaine L. 1996. Effects of fire and salvage logging on the cavity-nesting bird community in northwestern Montana. Missoula, MT: University of Montana. Dissertation. 115 p.

Chapman, H. H. 1912. Forest-fires and forestry in the southern states. American Forests. 18: 510-517.

Christensen, N. L. 1977. Fire and soil-plant nutrient relations in a pine-wiregrass savannah on the coastal plain of North Carolina. Oecologia. 31:27-44.

Christensen, Norman L. 1988. Succession and natural disturbance: paradigms, problems, and preservation of natural ecosystems. In: Agee, James K.; Johnson, Darryll R., eds. Ecosystem management for parks and wilderness. Seattle, WA: University of Washington Press: 62-86.

Christensen, Norman L.; Muller, C. H. 1975. Effects of fire on factors controlling growth in *Adenostoma* chaparral. Ecological Monographs. 45: 29-55.

Clark, J. S. 1988. Effect of climate change on fire regimes in northwestern Minnesota. Nature. 334: 233-235.

Coggins, Joe L.; Engle, J. W., Jr. 1971. Prescribed burning for blueberries. Virginia Wildlife. 22(8): 17-18.

Conant, Roger; Collins, Joseph T. 1991. A field guide to reptiles and amphibians, eastern and central North America, 3rd ed. Boston, MA: Houghton Mifflin Company. 450 p.

Cooper, C. F. 1960. Changes in vegetation, structure, and growth of southwestern pine forests since white settlement. Ecological Monographs. 30: 129-164.

Coppock, D. L.; Ellis, J. E.; Detling, J. K.; Dyer, M. I. 1983. Plant-herbivore interactions in a North American mixed-grass prairie. II. Responses of bison to modification of vegetation by prairie dogs. Oecologia. 56: 10-15.

Costa, R.; Escano, R. E. F. 1989. Red-cockaded woodpecker: status and management in the Southern Region in 1986. Technical report R8-TP12. Atlanta, GA: U.S. Department of Agriculture, Forest Service, Southern Region. 71 p.

Courtney, Rick F. 1989. Pronghorn use of recently burned mixed prairie in Alberta. Journal of Wildlife Management. 53(2): 302-305.

Crawford, John A. 1999. [personal communication]. November 1. Corvallis, OR: Oregon State University.

Croskery, P. R.; Lee, P. F. 1981. Preliminary investigations of regeneration patterns following wildfire in the boreal forest of northwestern Ontario. Alces. 17: 229-256.

Crowner, Ann W.; Barrett, Gary W. 1979. Effects of fire on the small mammal component of an experimental grassland community. Journal of Mammology. 60: 803-813.

Cumming, S. G.; Burton, P. J.; Klinkenberg, B. 1996. Boreal mixedwood forests may have no "representative" areas: some implications for reserve design. Ecography. 19: 162-180.

Daubenmire, R. 1968. Ecology of fire in grasslands. In: Cragg, J. B., ed. Advances in ecological research. Vol. 5. New York, NY: Academic Press: 209-266.

Davis, James L.; Valkenburg, Patrick. 1983. Calving in recently burned habitat by caribou displaced from their traditional calving area. Proceedings, Alaska Science Conference. 34: 19.

Day, G. M. 1953. The Indian as an ecological factor in the northeastern forest. Ecology. 34: 329-346.

DeBano, Leonard F.; Neary, Daniel G.; Ffolliott, Peter F. 1998. Fire's effects on ecosystems. New York, NY: John Wiley and Sons, Inc. 333 p.

DeByle, Norbert V.; Urness, Philip J.; Blank, Deborah L. 1989. Forage quality in burned and unburned aspen communities. Res. Pap. INT-404. Ogden, UT: U.S. Department of Agriculture, Forest Service, Intermountain Research Station. 8 p.

deMaynadier, P. G.; Hunter, M. L., Jr. 1995. The relationship between forest management and amphibian ecology: a review of the North American literature. Environmental Reviews. 3: 230-261.

Deming, O. V. 1963. Antelope and sagebrush. In: Yoakum, Jim, comp. Transactions, Interstate Antelope Conference; 1963 December 4-5; Alturas, CA. Reno, NV: Interstate Antelope Conference: 55-60.

DeWitt, James B.; Derby, James V., Jr. 1955. Changes in nutritive value of browse plants following forest fires. Journal of Wildlife Management. 19(1): 65-70.

Diaz, N.; Apostol, D. 1992. Forest landscape analysis and design: a process for developing and implementing land management objectives for landscape patterns. R6 ECO-TP-043-92. Portland, OR: U.S. Department of Agriculture, Forest Service, Pacific Northwest Region.

Dix, Ralph L.; Butler, John E. 1954. The effects of fire on a dry, thin-soil prairie in Wisconsin. Journal of Range Management. 7: 265-268.

Dodd, Norris L. 1988. Fire management and southwestern raptors. In: Glinski, R. L.; Pendleton, Beth Giron; Moss, Mary Beth; [and others], eds. Proceedings of the southwest raptor symposium and workshop; 1986 May 21-24; Tucson, AZ. NWF Scientific and Technology Series No. 11. Washington, DC: National Wildlife Federation: 341-347.

Drut, Martin A.; Pyle, William H.; Crawford, John A. 1994. Technical note: diets and food selection of sage grouse chicks in Oregon. Journal of Range Management. 47(1): 90-93.

Dunaway, Mervin Alton, Jr. 1976. An evaluation of unburned and recently burned longleaf pine forest for bobwhite quail brood habitat. Starkville, MS: Mississippi State University. Thesis. 32 p.

Duncan, B. A.; Boyle, S.; Breininger, D. R.; Schmalzer, P. A. 1999. Coupling past management practice and historical landscape change on John F. Kennedy Space Center. Landscape Ecology. 14: 291-309.

Ehrlich, Paul R.; Dobkin, David S.; Wheye, Darryl. 1988. The birder's handbook. New York, NY: Simon and Schuster Inc. 785 p.

Elliott, Bruce. 1985. Changes in distribution of owl species subsequent to habitat alteration by fire. Western Birds. 16(1): 25-28.

Emlen, John T. 1970. Habitat selection by birds following a forest fire. Ecology. 51(2): 343-345.

Engstrom, R. Todd; Crawford, Robert L.; Baker, W. Wilson. 1984. Breeding bird populations in relation to changing forest structure following fire exclusion: a 15-year study. Wilson Bulletin. 96(3): 437-450.

Erwin, William J.; Stasiak, Richard H. 1979. Vertebrate mortality during the burning of reestablished prairie in Nebraska. American Midland Naturalist. 101(1): 247-249.

Etchberger, Richard C. 1990. Effects of fire on desert bighorn sheep habitat. In: Krausman, Paul R.; Smith, Norman S., eds. Managing Wildlife in the Southwest Symposium: 53-57.

Evans, Raymond A.; Young, James A.; Cluff, Greg J.; McAdoo, J. Kent. 1983. Dynamics of antelope bitterbrush seed caches. In: Tiedemann, Arthur R.; Johnson, Kendall L., comps. Proceedings—research and management of bitterbrush and cliffrose in western North America; 1982 April 13-15; Salt Lake City, UT. Gen. Tech. Rep. INT-152. Ogden, UT: U.S. Department of Agriculture, Forest Service, Intermountain Forest and Range Experiment Station: 195-202.

Evans, Timothy L.; Guynn, David C., Jr.; Waldrop, Thomas A. 1991. Effects of fell-and-burn site preparation on wildlife habitat and small mammals in the upper southeastern piedmont. In: Nodvin, Stephen C.; Waldrop, Thomas A., eds. Fire and the environment: ecological and cultural perspectives: Proceedings of an international symposium; 1990 March 20-24; Knoxville, TN. Gen. Tech. Rep. SE-69. Asheville, NC: U.S. Department of Agriculture, Forest Service, Southeastern Forest Experiment Station: 160-167.

Evans, William G. 1971. The attraction of insects to forest fires. In: Proceedings, Tall Timbers conference on ecological animal control by habitat management; 1971 February 25-27; Tallahassee, FL. Number 3. Tallahassee, FL: Tall Timbers Research Station: 115-127.

Ewel, Katherine C. 1990. Swamps. In: Myers, Ronald L.; Ewel, John J., eds. Ecosystems of Florida. Orlando, FL: University of Central Florida Press: 281-322.

Eyre, F. H. 1980. Forest cover types of the United States and Canada. Washington, DC: Society of American Foresters. 148 p.

Ffolliott, Peter F.; Guertin, D. Phillip. 1990. Prescribed fire in Arizona ponderosa pine forests: a 24-year case study. In: Krammes, J. S., tech. coord. Effects of fire management of southwestern natural resources: Proceedings; 1988 Nov. 15-17; Tucson, AZ. Gen. Tech. Rep. RM-191. Fort Collins, CO: U.S. Department of Agriculture, Forest Service, Rocky Mountain Forest and Range Experiment Station: 250-254.

Fiedler, Carl E.; Arno, Stephen F.; Harrington, Michael G. 1998. Reintroducing fire in ponderosa pine-fir forests after a century of fire exclusion. In: Pruden, Teresa L.; Brennan, Leonard A., eds. Fire in ecosystem management: shifting the paradigm from suppression to prescription: Proceedings, 20th fire ecology conference; 1996 May 1-10; Boise, ID. Tallahassee, FL: Tall Timbers Research Station: 245-249.

Fiedler, Carl E.; Cully, Jack F., Jr. 1995. A silvicultural approach to develop Mexican spotted owl habitat in southwest forests. Western Journal of Applied Forestry. 10(4): 144-148.

Finch, Deborah M.; Ganey, Joseph L.; Yong, Wang; Kimball, Rebecca T.; Sallabanks, Rex. 1997. Effects and interactions of fire, logging, and grazing. In: Block, William M.; Finch, Deborah M., tech. eds. Songbird ecology in southwestern ponderosa pine forests: a literature review. Gen. Tech. Rep. RM-GTR-292. Fort Collins, CO: U.S. Department of Agriculture, Forest Service, Rocky Mountain Forest and Range Experiment Station: 103-136.

Fischer, Richard A.; Reese, Kerry P.; Connelly, John W. 1996. An investigation on fire effects within xeric sage grouse brood habitat. Journal or Range Management. 49(3): 194-198.

Fitzgerald, Susan M.; Tanner, George W. 1992. Avian community response to fire and mechanical shrub control in south Florida. Journal of Range Management. 45(4): 396-400.

Ford, William M.; Menzel, M. Alex; McGill, David W.; Laerm, Joshua; McCay, Timothy S. 1999. Effects of a community restoration fire on small mammals and herpetofauna in the southern Appalachians. Forest Ecology and Management. 114: 233-243.

Forman, Richard T. T.; Godron, Michael. 1986. Landscape ecology. New York, NY: John Wiley and Sons. 619 p.

Fox, J. F. 1983. Post-fire succession of small-mammal and bird communities. In: Wein, Ross W.; MacLean, David A., eds. The role of fire in northern circumpolar ecosystems. New York, NY: John Wiley and Sons: 155-180.

French, Marilynn Gibbs; French, Steven P. 1996. Large mammal mortality in the 1988 Yellowstone fires. In: Greenlee, Jason M., ed. The ecological implications of fire in Greater Yellowstone. Proceedings, 2nd biennial conference on the Greater Yellowstone Ecosystem; 1993 September 19-21; Yellowstone National Park, WY. Fairfield, WA: International Association of Wildland Fire: 113-115.

Frost, Cecil C. 1998. Presettlement fire frequency regimes of the United States: a first approximation. In: Pruden, Teresa L.;

USDA Forest Service Gen. Tech. Rep. RMRS-GTR-42-vol. 1. 2000

65

Brennan, Leonard A., eds. Fire in ecosystem management: shifting the paradigm from suppression to prescription: Proceedings, 20th Tall Timbers fire ecology conference; 1996 May 7-10; Tallahassee, FL. Tallahassee, FL: Tall Timbers Research Station: 70-81.

Frost, Cecil C.; Walker, J.; Peet, R. K. 1986. Fire-dependent savannahs and prairies of the southeast: Original extent, preservation status, and management problems. In: Kulhavy, D. L.; Conner, R. N., eds. Wilderness and natural areas in the eastern United States: a management challenge. Nacogdoches, TX: Stephen F. Austin State University, School of Forestry, Center for Applied Studies: 348-357.

Fule, Peter Z.; Covington, W. Wallace. 1995. Changes in fire regimes and forest structures of unharvested Petran and Madrean pine forests. In: DeBano, Leonard F.; Ffolliott, Peter F., Ortega-Rubio, Alfredo; Gottfried, Gerald J.; Hamre, Robert H.; Edminster, Carleton B., tech. coords. Biodiversity and management of the Madrean Archipelago; the sky islands of southwestern United States and northwestern Mexico: Proceedings; 1994 September 19-23; Tucson, AZ. Gen. Tech. Rep. RM-GTR-264. Fort Collins, CO: U.S. Department of Agriculture, Forest Service, Rocky Mountain Forest and Range Experiment Station: 408-415.

Fults, Gene A. 1991. Florida ranchers manage for deer. Rangelands. 13(1): 28-30.

Furyaev, V. V.; Wein, Ross W.; MacLean, David A. 1983. Fire influences in Abies-dominated forests. In: Wein, Ross W.; MacLean, David A., eds. The role of fire in northern circumpolar ecosystems. New York, NY: John Wiley and Sons: 221-234.

Gaines, William L.; Strand, Robert A.; Piper, Susan D. 1997. Effects of the Hatchery Complex fires on northern spotted owls in the eastern Washington Cascades. In: Greenlee, Jason M., ed. Proceedings, 1st conference on fire effects on rare and endangered species and habitats; 1995 November 13-16; Coeur d'Alene, ID. Fairfield, WA: International Association of Wildland Fire: 123-129.

Gasaway, W. C.; Dubois, S. D. 1985. Initial response of moose, *Alces alces*, to a wildfire in interior Alaska. Canadian Field-Naturalist. 99: 135-140.

Gasaway, W. C.; DuBois, S. D.; Boertje, R. D.; Reed, D. J.; Simpson, D. T. 1989. Response of radio-collared moose to a large burn in central Alaska. Canadian Journal of Zoology. 67(2): 325-239.

Geluso, Kenneth N.; Schroder, Gene D.; Bragg, Thomas B. 1986. Fire-avoidance behavior of meadow voles (*Microtus pennsylvanicus*). American Midland Naturalist. 116(1): 202-205.

Gill, A. M. 1998. An hierarchy of fire effects: impact of fire regimes on landscapes. In: Viegas, D. X., ed. Proceedings, Volume I, III International Conference on Forest Fire Research and 14th Conference on Fire and Forest Meteorology; 1998 November 16-20; Coimbra, Portugal. Coimbra, Portugal: ADAI - Associacao para o Desenvolvimento da Aerodinamica Industrial: 129-143.

Gilliam, Frank S. 1991. The significance of fire in an oligotrophic forest ecosystem. In: Nodvin, Stephen C.; Waldrop, Thomas A., eds. Fire and the environment: ecological and cultural perspectives: Proceedings of an international symposium; 1990 March 20-24; Knoxville, TN. Gen. Tech. Rep. SE-69. Asheville, NC: U.S. Department of Agriculture, Forest Service, Southeastern Forest Experiment Station: 113-122.

Graham, Russell T.; Harvey, Alan E.; Jurgensen, Martin F.; Jain, Theresa B.; Tonn, Jonalea R.; Page-Dumroese, Deborah S. 1994. Managing coarse woody debris in forests of the Rocky Mountains. Res. Pap. INT-RP-477. Ogden, UT: U.S. Department of Agriculture, Forest Service, Intermountain Research Station. 13 p.

Graham, Russell T; Jain, Theresa B.; Reynolds, Richard T.; Boyce, Douglas A. 1997. The role of fire in sustaining northern goshawk habitat in Rocky Mountain forests. In: Greenlee, Jason M., ed. Proceedings, 1st conference on fire effects on rare and endangered species and habitats; 1995 November 13-16; Coeur d'Alene, ID. Fairfield, WA: International Association of Wildland Fire: 69-76.

Graham, Russell T.; Rodriguez, Ronald L.; Paulin, Kathleen M.; Player, Rodney L.; Heap, Arlene P.; Williams, Richard. 1999. The northern goshawk in Utah: habitat assessment and management recommendations. Gen. Tech. Rep. RMRS-GTR-22. Ogden, UT: U.S. Department of Agriculture, Forest Service, Rocky Mountain Research Station. 48 p.

Grange, Wallace B. 1948. The relation of fire to grouse. In: Wisconsin grouse problems. Federal Aid in Wildlife Restoration Project No. 5R. Publication No. 328. Madison, WI: Wisconsin Conservation Department: 193-205.

Granholm, Stephen Lee. 1982. Effects of surface fires on birds and their habitat associations in coniferous forests of the Sierra Nevada, California. Davis, CA: University of California. Dissertation. 60 p.

Green, D. G. 1986. Pollen evidence for the postglacial origins of Nova Scotia's forests. Canadian Journal of Botany. 65: 1163-1179.

Green, S. W. 1931. The forest that fire made. American Forests. 37: 583-584, 618.

Gregg, Michael A.; Crawford, John A.; Drut, Martin S.; DeLong, Anita K. 1994. Vegetational cover and predation of sage grouse nests in Oregon. Journal of Wildlife Management. 58(1): 162-166.

Groves, Craig R.; Steenhof, Karen. 1988. Responses of small mammals and vegetation to wildfire in shadscale communities of southwestern Idaho. Northwest Science. 62(5): 205-210.

Gruell, George E. 1979. Wildlife habitat investigations and management implications on the Bridger-Teton National Forest. In: Boyce, Mark S.; Hayden-Wing, Larry D., eds. North American elk, ecology, behavior and management. Laramie, WY: University of Wyoming: 63-74.

Gruell, George E.; Schmidt, Wyman C.; Arno, Stephen F.; Reich, William J. 1982. Seventy years of vegetative change in a managed ponderosa pine forest in western Montana—implications for resource management. Gen. Tech. Rep. INT-130. U.S. Department of Agriculture, Forest Service, Intermountain Forest and Range Experiment Station. 42 p.

Habeck, J. R.; Mutch, R. W. 1973. Fire-dependent forests in the northern Rocky Mountains. Quaternary Research. 3: 408-424.

Hallisey, Dennis M.; Wood, Gene W. 1976. Prescribed fire in scrub oak habitat in central Pennsylvania. Journal of Wildlife Management. 40(3): 507-516.

Hamas, Michael J. 1983. Nest-site selection by eastern kingbirds in a burned forest. Wilson Bulletin. 95(3): 475-477.

Hamer, David. 1995. Buffaloberry (*Shepherdia canadensis*) fruit production in fire-successional bear feeding sites. Report to Parks Canada, Banff National Park. Banff, AB: Parks Canada, Banff National Park. 65 p.

Hamilton, Evelyn H.; Yearsley, H. Karen. 1988. Vegetation development after clearcutting and site preparation in the SBS zone. Economic and Regional Development Agreement: FRDA Report 018, ISSN 0835 0752. Victoria, BC: Canadian Forestry Service, Pacific Forestry Centre; British Columbia Ministry of Forests and Lands, Research Branch. 66 p.

Hamilton, Robert J. 1981. Effects of prescribed fire on black bear populations in the southern forests. In: Wood, Gene W., ed. Prescribed fire and wildlife in southern forests: Proceedings of a symposium; 1981 April 6-8; Myrtle Beach, SC. Georgetown, SC: The Belle W. Baruch Forest Science Institute of Clemson University: 129-134.

Harlow, R. F.; Van Lear, D. H. 1989. Effects of prescribed burning on mast production in the Southeast. In: McGee, C. E., ed. Southern Appalachian mast management: Proceedings of the workshop; 1989 August 14-16; Knoxville, TN. Knoxville, TN: University of Tennessee, Department of Forestry, Wildlife and Fisheries; U.S. Department of Agriculture, Forest Service: 54-65.

Harrington, Michael G. 1996. Fall rates of prescribed fire-killed ponderosa pine. Res. Pap. INT-RP-489. Ogden, UT: U.S. Department of Agriculture, Forest Service, Intermountain Research Station. 7 p.

Hart, Stephen. 1998. Beetle mania: an attraction to fire. BioScience. 48(1): 3-5.

Hedlund, J. D.; Rickard, W. H. 1981. Wildfire and the short-term response of small mammals inhabiting a sagebrush-bunchgrass community. Murrelet. 62(1): 10-14.

Heinselman, M. L. 1981. Fire intensity and frequency as factors in the distribution and structure of northern ecosystems. In: Mooney, H. A.; Bonnicksen, T. M.; Christensen, N. L.; Lotan, J. E.; Reiners, W. A., tech. coords. Fire regimes and ecosystem properties: Proceedings of the conference; 1978 December 11-15; Honolulu, HI. Gen. Tech. Rep. WO-26. Washington, DC: U.S. Department of Agriculture, Forest Service: 7-57.

66

USDA Forest Service Gen. Tech. Rep. RMRS-GTR-42-vol. 1. 2000

Heinselman, Miron L. 1973. Fire in the virgin forests of the Boundary Waters Canoe Area, Minnesota. Journal of Quaternary Research. 3: 329-382.

Heinselman, Miron L. 1978. Fire in wilderness ecosystems. In: Hendee, John C.; Stankey, George H.; Lucas, Robert C. Wilderness management. Misc. Pub. No. 1365. Washington, DC: U.S. Department of Agriculture, Forest Service: 249-278.

Hejl, Sallie; McFadzen, Mary. 1998. Interim report: Maintaining fire-associated bird species across forest landscapes in the northern Rockies. Unpublished report on file at: U.S. Department of Agriculture, Forest Service, Rocky Mountain Research Station, Fire Sciences Laboratory, Missoula, MT. 15 p.

Helms, John A., ed. 1998. The dictionary of forestry. Bethesda, MD: Society of American Foresters. 210 p.

Higgins, K. F. 1984. Lightning fires in North Dakota grasslands and in pine-savanna lands of South Dakota and Montana. Journal of Range Management. 37: 100-103.

Higgins, K. F. 1986. A comparison of burn season effects on nesting birds in North Dakota mixed-grass prairie. Prairie Naturalist. 18(4): 219-228.

Hilmon, J. B.; Hughes, R. H. 1965. Fire and forage in the wiregrass type. Journal of Range Management. 18: 251-254.

Hines, William W. 1973. Black-tailed deer populations and Douglas-fir reforestation in the Tillamook Burn, Oregon. Game Research Report Number 3. Federal Aid to Wildlife Restoration, Project W-51-R: Final Report. Corvallis, OR: Oregon State Game Commission. 59 p.

Hobbs, N. T.; Spowart, R. A. 1984. Effects of prescribed fire on nutrition of mountain sheep and mule deer during winter and spring. Journal of Wildlife Management. 48(2): 551-560.

Hobbs, N. Thompson; Schimel, David S. 1984. Fire effects on nitrogen mineralization and fixation in mountain shrub and grassland communities. Journal of Range Management. 37(5): 402-405.

Hobbs, N. Thompson; Schimel, David S.; Owensby, Clenton E.; Ojima, Dennis S. 1991. Fire and grazing in the tallgrass prairie: contingent effects on nitrogen budgets. Ecology. 72(4): 1374-1382.

Horton, Scott P.; Mannan, R. William. 1988. Effects of prescribed fire on snags and cavity-nesting birds in southeastern Arizona pine forests. Wildlife Society Bulletin. 16(1): 37-44.

Howard, W. E.; Fenner, R. L.; Childs, H. E., Jr. 1959. Wildlife survival in brush burns. Journal of Range Management. 12: 230-234.

Huber, G. E.; Steuter, A. A. 1984. Vegetation profile and grassland bird response to spring burning. Prairie Naturalist. 16(2): 55-61.

Huff, M. F. 1984. Post-fire succession in the Olympic Mountains, Washington: forest vegetation, fuels, and avifuana. Seattle, WA: University of Washington. Dissertation. 240 p.

Huff, Mark H.; Agee, James K; Manuwal, David A. 1985. Postfire succession of avifauna in the Olympic Mountains, Washington. In: Lotan, James E.; Brown, James K., comps. Fire's effects on wildlife habitat—symposium proceedings; 1984 March 21; Missoula, MT. Gen. Tech. Rep. INT-186. Ogden, UT: U.S. Department of Agriculture, Forest Service, Intermountain Research Station: 8-15.

Hulbert, Lloyd C. 1986. Fire effects on tallgrass prairie. In: Clambey, Gary K.; Pemble, Richard H., eds. The prairie: past, present and future: Proceedings, 9th North American prairie conference; 1984 July 29-August 1; Moorhead, MN. Fargo, ND: Tri-College University, Center for Environmental Studies: 138-142.

Hulbert, Lloyd C. 1988. Causes of fire effects in tallgrass prairie. Ecology. 69(1): 46-58.

Humphrey, Robert R. 1974. Fire in the deserts and desert grassland. In: Kozlowski, T. T.; Ahlgren, C. E., eds. Fire and ecosystems. New York, NY: Academic Press: 365-400.

Hurst, George A. 1971. The effects of controlled burning on arthropod density and biomass in relation to bobwhite quail brood habitat on a right-of-way. In: Proceedings, Tall Timbers conference on ecological animal control by habitat management; 1970 February 26-28; Tallahassee, FL. Number 2. Tallahassee, FL: Tall Timbers Research Station: 173-183.

Hurst, George A. 1978. Effects of controlled burning on wild turkey poult food habits. Proceedings, Annual Conference of the Southeastern Association of Fish and Wildlife Agencies. 32: 30-37.

Hutto, Richard L. 1995. Composition of bird communities following stand-replacement fires in northern Rocky Mountain conifer forests. Conservation Biology. 9(5): 1041-1058.

Irwin, Larry L. 1985. Foods of moose, *Alces alces*, and white-tailed deer, *Odocoileus virginianus*, on a burn in boreal forest. Canadian Field-Naturalist. 99(2): 240-245.

Irwin, Larry L.; Peek, James M. 1983. Elk habitat use relative to forest succession in Idaho. Journal of Wildlife Management. 47(3): 664-672.

Ivey, T. L.; Causey, M. K. 1984. Response of white-tailed deer to prescribed fire. Wildlife Society Bulletin. 12(2): 138-141.

James, T. D. W.; Smith, D. W. 1977. Short-term effects of surface fire on the biomass and nutrient standing crop of *Populus tremuloides* in southern Ontario. Canadian Journal of Forest Research. 7: 666-679.

Jameson, Donald A. 1962. Effects of burning on a galleta-black grama range invaded by juniper. Ecology. 43(4): 760-763.

Johnson, A. Sydney; Landers, J. Larry. 1978. Fruit production in slash pine plantations in Georgia. Journal of Wildlife Management. 42(3): 606-613.

Johnson, Edward A. 1992. Fire and vegetation dynamics: Studies from the North American boreal forest. Cambridge, UK: Cambridge University Press. 129 p.

Jones, J. Knox, Jr.; Horrmann, Robert S.; Rice, Dale W.; Jones, Clyde; Baker, Robert J.; Engstrom, Mark D. 1992. Revised checklist of North American mammals north of Mexico, 1991. Occasional Paper Number 146. Lubbock, TX: The Museum, Texas Tech University. 23 p.

Jourdonnais, C. S.; Bedunah, D. J. 1990. Prescribed fire and cattle grazing on an elk winter range in Montana. Wildlife Society Bulletin. 18: 232-240.

Kauffman, J. Boone; Martin, Robert E. 1985. Shrub and hardwood response to prescribed burning with varying season, weather, and fuel moisture. In: Proceedings, 8th conference on fire and forest meteorology; 1985 April 29-May 2; Detroit, MI. Bethesda, MD: Society of American Foresters: 279-286.

Kaufman, Donald W.; Gurtz, Sharon K.; Kaufman, Glennis A. 1988a. Movements of the deer mouse in response to prairie fire. Prairie Naturalist. 20(4): 225-229.

Kaufman, Glennis A.; Kaufman, Donald W.; Finck, Elmer J. 1982. The effect of fire frequency on populations of the deer mouse (*Peromyscus maniculatus*) and the western harvest mouse (*Reithrodontomys megalotis*). Bulletin of the Ecological Society of America. 63: 66.

Kaufman, Glennis A.; Kaufman, Donald W.; Finck, Elmer J. 1988b. Influence of fire and topography on habitat selection by *Peromyscus maniculatus* and *Reithrodontomys megalotis* in ungrazed tallgrass prairie. Journal of Mammology. 69(2): 342-352.

Kay, Charles E. 1995. Aboriginal overkill and native burning: implications for modern ecosystem management. Western Journal of Applied Forestry. 10(4): 121-126.

Kay, Charles E. 1998. Are ecosystems structured from the top-down or bottom-up: a new look at an old debate. Wildlife Society Bulletin. 26(3): 484-498.

Keay, Jeffrey A.; Peek, James M. 1980. Relationships between fires and winter habitat of deer in Idaho. Journal of Wildlife Management. 44(2): 372-380.

Kelleyhouse, David G. 1979. Fire/wildlife relationships in Alaska. In: Hoefs, M.; Russell, D., eds. Wildlife and wildfire: Proceedings of workshop; 1979 November 27-28; Whitehorse, YT. Whitehorse, YT: Yukon Wildlife Branch: 1-37.

Kilgore, B. M. 1976. From fire control to management: an ecological basis for policies. Transactions, North American Wildlife and Natural Resources Conferences. 41: 477-493.

Kilgore, B. M. 1981. Fire in ecosystem distribution and structure: western forests and scrublands. In: Mooney, H. A.; Bonnicksen, T. M.; Christensen, N. L.; Lotan, J. E.; Reiners, W. A., tech. coords. Proceedings of the conference: fire regimes and ecosystem properties; 1978 December 11-15; Honolulu, HI. Gen. Tech. Rep. WO-26. Washington, DC: U.S. Department of Agriculture, Forest Service: 58-89.

Kirsch, Leo M. 1974. Habitat management considerations for prairie chickens. Wildlife Society Bulletin. 2(3): 124-129.

Klebenow, Donald A. 1969. Sage grouse nesting and brood habitat in Idaho. Journal of Wildlife Management. 33(3): 649-662.

USDA Forest Service Gen. Tech. Rep. RMRS-GTR-42-vol. 1. 2000

67

Klebenow, Donald A. 1973. The habitat requirements of sage grouse and the role of fire in management. In: Proceedings, 12th annual Tall Timbers fire ecology conference; 1972 June 8-9; Lubbock, TX. Tallahassee, FL: Tall Timbers Research Station: 305-315.

Klein, David R. 1982. Fire, lichens, and caribou. Journal of Range Management. 35(3): 390-395.

Klinger, R. C.; Kutilek, M. J.; Shellhammer, H. S. 1989. Population responses of black-tailed deer to prescribed burning. Journal of Wildlife Management. 53: 863-871.

Knick, Steven T. 1999. Requiem for a sagebrush ecosystem? Northwest Science. 73(1): 53-57.

Knick, Steven T.; Rotenberry, John T. 1995. Landscape characteristics of fragmented shrubsteppe habitats and breeding passerine birds. Conservation Biology. 9(5): 1059-1071.

Knodel-Montz, J. J. 1981. Use of artificial perches on burned and unburned tallgrass prairie. Wilson Bulletin. 93(4): 547-548.

Kobriger, Jerry D.; Vollink, David P.; McNeill, Michael E.; Higgins, Kenneth F. 1988. Prairie chicken populations of the Sheyenne Delta in North Dakota, 1961-1987. In: Bjugstad, Ardell J., tech. coord. Prairie chickens on the Sheyenne National Grasslands— symposium proceedings; 1987 September 18; Crookston, MN. Gen. Tech. Rep. RM-159. Fort Collins, CO: U.S. Department of Agriculture, Forest Service, Rocky Mountain Forest and Range Experiment Station: 1-7.

Koehler, Gary M.; Aubry, Keith B. 1994. Lynx. In: Ruggiero, Leonard F.; Aubry, Keith B.; Buskirk, Steven W.; Lyon, L. Jack; Zielinski, William J., tech. eds. American marten, fisher, lynx, and wolverine in the western United States: The scientific basis for conserving forest carnivores. Gen. Tech. Rep. RM-254. Fort Collins, CO: U.S. Department of Agriculture, Forest Service, Rocky Mountain Forest and Range Experiment Station: 74-98.

Koehler, Gary M.; Hornocker, Maurice G. 1977. Fire effects on marten habitat in the Selway-Bitterroot Wilderness. Journal of Wildlife Management. 41(3): 500-505.

Koerth, Ben H.; Mutz, James L.; Segers, James C. 1986. Availability of bobwhite foods after burning of Pan American balsamscale. Wildlife Society Bulletin. 14(2): 146-150.

Komarek, E. V., Sr. 1962. The use of fire: an historical background. In: Proceedings, 1st annual Tall Timbers fire ecology conference; 1962 March 1-2; Tallahassee, FL. Tallahassee, FL: Tall Timbers Research Station: 7-10.

Komarek, E. V., Sr. 1968. Lightning and lightning fires as ecological forces. In: Proceedings, 8th annual Tall Timbers fire ecology conference; 1968 March 14-15; Tallahassee, FL. Tallahassee, FL: Tall Timbers Research Station: 169-197.

Komarek, E. V., Sr. 1969. Fire and animal behavior. In: Proceedings, 9th annual Tall Timbers Fire Ecology conference; 1969 April 10-11; Tallahassee, FL. Tallahassee, FL: Tall Timbers Research Station: 161-207.

Komarek, E. V. 1974. Effects of fire on temperate forests and related ecosystems: southeastern United States. In: Kozlowski, T. T.; Ahlgren, C. E., eds. 1974. Fire and ecosystems. New York, NY: Academic Press: 251-277.

Koniak, Susan. 1985. Succession in pinyon-juniper woodlands following wildfire in the Great Basin. Great Basin Naturalist. 45(3): 556-566.

Kramp, Betty A.; Patton, David R.; Brady, Ward W. 1983. RUN WILD: Wildlife/habitat relationships: The effects of fire on wildlife habitat and species. Albuquerque, NM: U.S. Department of Agriculture, Forest Service, Southwestern Region, Wildlife Unit Technical Report. 29 p.

Kruse, Arnold D.; Piehl, James L. 1986. The impact of prescribed burning on ground-nesting birds. In: Clambey, Gary K.; Pemble, Richard H., eds. The prairie: past, present and future: Proceedings, 9th North American prairie conference; 1984 July 29-August 1; Moorhead, MN. Fargo, ND: Tri-College University Center for Environmental Studies: 153-156.

Kucera, Clair L. 1981. Grasslands and fire. In: Mooney, H. A.; Bonnicksen, T. M.; Christensen, N. L.; Lotan, J. E.; Reiners, W. A., tech. coords. Proceedings of the conference: Fire regimes and ecosystem properties; 1978 December 11-15; Honolulu, HI. Gen. Tech. Rep. WO-26. Washington, DC: U.S. Department of Agriculture, Forest Service: 90-111.

Kufeld, Roland C. 1983. Responses of elk, mule deer, cattle, and vegetation to burning, spraying, and chaining of Gambel oak

rangeland. Tech. Pub. DOW-R-T-34-'83. Fort Collins, CO: Colorado Division of Wildlife, Research Center Library. 47 p.

Kwilosz, John R.; Knutson, Randy L. 1999. Prescribed fire management of Karner blue butterfly habitat at Indiana Dunes National Lakeshore. Natural Areas Journal. 19(2): 98-108.

Landers, J. Larry. 1987. Prescribed burning for managing wildlife in southeastern pine forests. In: Dickson, James G.; Maughan, O. Eugene, eds. Managing southern forests for wildlife and fish: a proceedings; 1986 October 5-8; Birmingham, AL. Gen. Tech. Rep. SO-65. New Orleans, LA: U.S. Department of Agriculture, Forest Service, Southern Forest Experiment Station: 19-27.

Launchbaugh, J. L. 1972. Effect of fire on shortgrass and mixed prairie species. In: Proceedings, 12th annual Tall Timbers fire ecology conference; 1972 June 8-9; Lubbock, TX. Tallahassee, FL: Tall Timbers Research Station: 129-151.

Lawrence, G. E. 1966. Ecology of vertebrate animals in relation to chaparral fire in the Sierra Nevada foothills. Ecology. 47(2): 278-290.

Laymon, Stephen A. 1985. General habitats and movements of spotted owls in the Sierra Nevada. In: Gutierrez, Ralph J.; Carey, Andrew B., tech. eds. Ecology and management of the spotted owl in the Pacific Northwest; 1984 June 19-23; Arcata, CA. Gen. Tech. Rep. PNW-185. Portland, OR: U.S. Department of Agriculture, Forest Service, Pacific Northwest Forest and Range Experiment Station: 66-68.

Leege, Thomas A.; Hickey, William O. 1971. Sprouting of northern Idaho shrubs after prescribed burning. Journal of Wildlife Management. 35(3): 508-515.

Lehman, Robert N.; Allendorf, John W. 1989. The effects of fire, fire exclusion and fire management on raptor habitats in the western United States. Western Raptor Management Symposium and Workshop: 236-244.

Leopold, A. S.; Cain, S. A.; Cottam, C. M.; Gabrielson, I. N.; Kimball, T. L. 1963. Wildlife management in the National Parks. In: Administrative policies for natural areas of the National Park system. Washington, DC: U.S. Department of the Interior, National Park Service. 14 p.

Lertzman, Ken; Fall, Joseph; Dorner, Brigitte. 1998. Three kinds of heterogeneity in fire regimes: at the crossroads of fire history and landscape ecology. Northwest Science. 72(special issue): 4-23.

Lillywhite, H. B.; North, F. 1974. Perching behavior of Sceloporus occidentalis in recently burned chaparral. Copeia. 1974: 256-257.

Lincoln, Roger; Boxshall, Geoff; Clark, Paul. 1998. A dictionary of ecology, evolution and systematics. 2nd ed. Cambridge, UK: Cambridge University Press. 361 p.

Little, S. 1974. Effects of fire on temperate forests: northeastern United States. In: Kozlowski, T. T.; Ahlgren, C. E., eds. Fire and ecosystems. New York, NY: Academic Press: 251-277.

Lloyd, Hoyes. 1938. Forest fire and wildlife. Journal of Forestry. 36: 1051-1054.

Loeb, S. C.; Pepper, W. D.; Doyle, A. T. 1992. Habitat characteristics of active and abandoned red-cockaded woodpecker colonies. Southern Journal of Applied Forestry. 16: 120-125.

Longland, William S. 1994. Seed use by desert granivores. In: Monsen, Stephen B.; Kitchen, Stanley G., comps. Proceedings— ecology and management of annual rangelands; 1992 May 18-21; Boise, ID. Gen. Tech. Rep. INT-GTR-313. Ogden, UT: U.S. Department of Agriculture, Forest Service, Intermountain Research Station: 233-237.

Longland, William S. 1995. Desert rodents in disturbed shrub communities and their effects on plant recruitment. In: Roundy, Bruce A.; McArthur, E. Durant; Haley, Jennifer S.; Mann, David K., comps. Proceedings: wildland shrub and arid land restoration symposium; 1993 October 19-21; Las Vegas, NV. Gen. Tech. Rep. INT-GTR-315. Ogden, UT: U.S. Department of Agriculture, Forest Service, Intermountain Research Station: 209-215.

Loveless, C. M. 1959. A study of the vegetation in the Florida Everglades. Ecology. 40: 1-9.

Lowe, P. O.; Ffolliott, P. F.; Dieterich, J. H.; Patton, D. R. 1978. Determining potential wildlife benefits from wildfire in Arizona ponderosa pine forests. Gen. Tech. Rep. RM-52. Fort Collins, CO: U.S. Department of Agriculture, Forest Service, Rocky Mountain Forest and Range Experiment Station. 12 p.

Lutz, H. J. 1956. Ecological effects of forest fires in the interior of Alaska. Tech. Bull. 1133. Washington, DC: U.S. Department of Agriculture. 121 p.

68

USDA Forest Service Gen. Tech. Rep. RMRS-GTR-42-vol. 1. 2000

Lyon, L. J. 1971. Vegetal development following prescribed burning of Douglas-fir in south-central Idaho. Res. Pap. INT-105. Ogden, UT: U.S. Department of Agriculture, Forest Service, Intermountain Forest and Range Experiment Station. 30 p.

Lyon, L. Jack; Crawford, Hewlette S.; Czuhai, Eugene; Frederiksen, R. L.; Harlow, R. F.; Metz, L. J.; Pearson, H. A. 1978. Effects of fire on fauna: a state of knowledge review. Gen. Tech. Report WO-6. Washington, DC: U.S. Department of Agriculture, Forest Service. 22 p.

Lyon, L. Jack; Marzluff, John M. 1985. Fire effects on a small bird population. In: Lotan, James E.; Brown, James K., comps. Fire's effects on wildlife habitat—symposium proceedings; 1984 March 21; Missoula, MT. Gen. Tech. Rep. INT-186. Ogden, UT: U.S. Department of Agriculture, Forest Service, Intermountain Research Station: 16-22.

MacCleery, D. W. 1993. American forests—a history of resiliency and recovery, revised edition. Durham, NC: U.S. Department of Agriculture, Forest Service and the Forest History Society. 58 p.

MacCracken, James G.; Viereck, Leslie A. 1990. Browse regrowth and use by moose after fire in interior Alaska. Northwest Science. 64(1): 11-18.

MacPhee, Douglas T. 1991. Prescribed burning and managed grazing restores tobosa grassland, antelope populations (Arizona). Restoration and Management Notes. 9(1): 35-36.

Martin, Robert C. 1990. Sage grouse responses to wildfire in spring and summer habitats. Moscow, ID: University of Idaho. Thesis. 36 p.

Mason, Robert B. 1981. Response of birds and rodents to controlled burning in pinyon-juniper woodlands. Reno, NV: University of Nevada. Thesis. 55 p.

Masters, R. E.; Skeen, J. E.; Whitehead, J. 1995. Preliminary fire history of McCurtain County Wilderness Area and implications for red-cockaded woodpecker management. In: Kulhavy, D. L.; Hooper, R. G.; Costa, R., eds. Red-cockaded woodpecker: recovery, ecology and management. Nacogdoches, TX: Stephen F. Austin State University, School of Forestry, Center for Applied Studies: 290-302.

Masters, Ronald E.; Lochmiller, Robert L.; Engle, David M. 1993. Effects of timber harvest and prescribed fire on white-tailed deer forage production. Wildlife Society Bulletin. 21(4): 401-411.

Mayfield, Harold F. 1963. Establishment of preserves for the Kirtland's warbler in the state and national forests of Michigan. Wilson Bulletin. 75(2): 216-220.

Mayfield, Harold F. 1993. Kirtland's Warblers benefit from large forest tracts. Wilson Bulletin. 105(2): 351-353.

Mazaika, Rosemary; Krausman, Paul R.; Etchberger, Richard C. 1992. Forage availability for mountain sheep in Pusch Ridge Wilderness, Arizona. Southwestern Naturalist. 37(4): 372-378.

McClelland, B. Riley. 1977. Relationships between hole-nesting birds, forest snags, and decay in western larch-Douglas-fir forests of the northern Rocky Mountains. Missoula, MT: University of Montana. Dissertation. 495 p.

McClure, H. Elliott. 1981. Some responses of resident animals to the effects of fire in a coastal chaparral environment in southern California. Cal-Neva Wildlife Transactions. 1981: 86-99.

McCoy, E. D.; Kaiser, B. W. 1990. Changes in foraging activity of the southern harvester ant Pogonomyrmex badius (Latreille) in response to fire. American Midland Naturalist. 123: 112-113.

McCulloch, C. Y. 1969. Some effects of wildfire on deer habitat in pinyon-juniper woodland. Journal of Wildlife Management. 33: 778-784.

McDonald, Philip M.; Huber, Dean W. 1995. California's hardwood resource: managing for wildlife, water, pleasing scenery, and wood products. Gen. Tech. Rep. PSW-GTR-154. Albany, CA: U.S. Department of Agriculture, Forest Service, Pacific Southwest Research Station. 23 p.

McIntosh, Robert P. 1985. The background of ecology: concept and theory. New York, NY: Cambridge University Press. 383 p.

McMahon, Thomas E.; deCalesta, David S. 1990. Effects of fire on fish and wildlife. In: Walstad, John D.; Radosevich, Steven R.; Sandberg, David V., eds. Natural and prescribed fire in Pacific Northwest forests. Corvallis, OR: Oregon State University Press: 233-250.

Means, D. Bruce; Campbell, Howard W. 1981. Effects of prescribed burning on amphibians and reptiles. In: Wood, Gene W., ed. Prescribed fire and wildlife in southern forests: Proceedings of a symposium; 1981 Apr 6-8; Myrtle Beach, SC. Georgetown, SC: The Belle W. Baruch Forest Science Institute of Clemson University: 89-96.

Meneely, S. C.; Schemnitz, S. C. 1981. Chemical composition and in vitro digestibility of deer browse three years after a wildfire. Southwestern Naturalist. 26:365-374.

Metz, Louis J.; Farrier, Maurice H. 1971. Prescribed burning and soil mesofauna on the Santee Experimental Forest. In: Prescribed burning symposium: Proceedings; 1971 April 14-16; Charleston, SC. Asheville, NC: U.S. Department of Agriculture, Forest Service, Southeastern Forest Experiment Station: 100-106.

Mihuc, Timothy B.; Minshall, G. Wayne; Robinson, Christopher T. 1996. Response of benthic macroinvertebrate populations in Cache Creek, Yellowstone National Park to the 1988 wildfires. In: Greenlee, Jason M., ed. The ecological implications of fire in Greater Yellowstone. Proceedings, 2nd biennial conference on the Greater Yellowstone Ecosystem; 1993 September 19-21; Yellowstone National Park, WY. Fairfield, WA: International Association of Wildland Fire: 83-94.

Miller, D. R. 1976. Taiga winter range relationships and diet. Canadian Wildlife Service Report Series No. 36. Ottawa, ON: Environment Canada, Wildlife Service. 42 p.

Minshall, G. W.; Brock, J. T.; Varley, J. D. 1989. Wildfires and Yellowstone's stream ecosystems. BioScience. 39: 707-715.

Moore, Conrad Taylor. 1972. Man and fire in the central North American grassland 1535-1890: a documentary historical geography. Los Angeles, CA: University of California. Dissertation. 155 p.

Moore, W. R. 1974. From fire control to management. Western Wildlands. 1: 11-15.

Morgan, Penelope; Aplet, Gregory H.; Haufler, Jonathan B.; Humphries, Hope C.; Moore, Margaret M.; Wilson, W. Dale. 1994. Historical range of variability: a useful tool for evaluating ecosystem change. Journal of Sustainable Forestry. 2(1/2): 87-112.

Morgan, Penelope; Bunting, Stephen C.; Keane, Robert E.; Arno, Stephen F. 1994. Fire ecology of whitebark pine forests of the northern Rocky Mountains, U.S.A. In: Schmidt, Wyman C.; Holtmeier, Friedrich-Karl, comps. Proceedings—International Workshop on Subalpine Stone Pines and Their Environment: the Status of Our Knowledge; 1992 September 5-11; St. Moritz, Switzerland. Gen. Tech. Rep. INT-GTR-309. Ogden, UT: U.S. Department of Agriculture, Forest Service, Intermountain Research Station: 136-141.

Moriarty, David J.; Farris, Richard E.; Noda, Diane K.; Stanton, Patricia A. 1985. Effects of fire on a coastal sage scrub bird community. Southwestern Naturalist. 30(3): 452-453.

Murphy, P. J. 1985. Methods for evaluating the effects of forest fire management in Alberta. Vancouver, BC: University of British Columbia. Dissertation. 160 p.

Murray, Michael P.; Bunting, Stephen C.; Morgan, Penny. 1997. Subalpine ecosystems: the roles of whitebark pine and fire. In: Greenlee, Jason M., ed. Proceedings, 1st conference on fire effects on rare and endangered species and habitats; 1995 November 13-16; Coeur d'Alene, ID. Fairfield, WA: International Association of Wildland Fire: 295-299.

Mushinsky, H. R.; Gibson, D. J. 1991. The influence of fire periodicity on habitat structure. In: Bell, S. S.; McCoy, E. D.; Mushinsky, H. R., eds. Habitat structure: the physical arrangement of objects in space. New York, NY: Chapman and Hall: 237-259.

Mutch, Robert W.; Arno, Stephen F.; Brown, James K.; Carlson, Clinton E.; Ottmar, Roger D.; Peterson, Janice L. 1993. Forest health in the Blue Mountains: a management strategy for fire-adapted ecosystems. Gen. Tech. Rep. PNW-GTR-310. Portland, OR: U.S. Department of Agriculture, Forest Service, Pacific Northwest Research Station. 14 p.

Myers, R. L. 1990. Scrub and high pine. In: Myers, R. L.; Ewel, J. J., eds. Ecosystems of Florida. Orlando, FL: University of Central Florida Press: 150-153.

National Park Service; USDA Forest Service; Bureau of Indian Affairs; U.S. Fish and Wildlife Service; Bureau of Land Management. 1998. Wildland prescribed fire management policy: Implementation procedures reference guide. Boise, ID: U.S. Department of the Interior, National Park Service, National Interagency Fire Center. 78 p.

USDA Forest Service Gen. Tech. Rep. RMRS-GTR-42-vol. 1. 2000

69

Nichols, R.; Menke, J. 1984. Effects of chaparral shrubland fire on terrestrial wildlife. In: DeVries, Johannes J., ed. Shrublands in California: literature review and research needed for management. Contribution No. 191, ISSN 0575-4941. Davis, CA: University of California, California Water Resources Center: 74-97.

O'Halloran, Kathleen A.; Blair, Robert M.; Alcaniz, Rene; Morris, Hershel F., Jr. 1987. Prescribed burning effects on production and nutrient composition of fleshy fungi. Journal of Wildlife Management. 51(1): 258-262.

O'Hara, Kevin L.; Latham, Penelope A.; Hessburg, Paul; Smith, Bradley G. 1996. A structural classification for Inland Northwest vegetation. Western Journal of Applied Forestry. 11(3): 97-102.

Ohmann, Lewis F.; Grigal, David F. 1979. Early revegetation and nutrient dynamics following the 1971 Little Sioux forest fire in northeastern Minnesota. Forest Science Monograph 21. Washington, DC: Society of American Foresters. 80 p.

Oldemeyer, J. L.; Franzmann, A. W.; Brundage, A. L.; Arneson, P. D.; Flynn, A. 1977. Browse quality and the Kenai moose population. Journal of Wildlife Management. 41(3): 533-542.

Oliver, Chadwick D.; Osawa, Akira; Camp, Ann. 1998. Forest dynamics and resulting animal and plant population changes at the stand and landscape levels. Journal of Sustainable Forestry. 6(3/4): 281-312.

Oswald, Brian P.; Covington, W. Wallace. 1983. Changes in understory production following a wildfire in southwestern ponderosa pine. Journal of Range Management. 36(4): 507-509.

Pack, J. C.; Williams, K. I.; Taylor, C. I. 1988. Use of prescribed burning in conjunction with thinning to increase wild turkey brood range habitat in oak-hickory forests. Transactions, Northeast Section of the Wildlife Society. 45: 37-48.

Parker, J. W. 1974. Activity of red-tailed hawks at a corn stubble fire. Kansas Ornithological Society. 22: 17-18.

Payette, S.; Morneau, C.; Sirios, L.; Desponts, M. 1989. Recent fire history in northern Québec biomes. Ecology. 70: 656-673.

Pearson, H. A.; Davis, J. R.; Schubert, G. H. 1972. Effects of wildfire on timber and forage production in Arizona. Journal of Range Management. 25: 250-253.

Peck, V. Ross; Peek, James M. 1991. Elk, Cervus elaphus, habitat use related to prescribed fire, Tuchodi River, British Columbia. Canadian Field-Naturalist. 105: 354-362.

Peek, James M. 1972. Adaptations to the burn: moose and deer studies. Minnesota Naturalist. 23(3-4): 8-14.

Peek, James M. 1974. Initial response of moose to a forest fire in northeastern Minnesota. American Midland Naturalist. 91(2): 435-438.

Peek, James. M.; Demarchi, Dennis A.; Demarchi, Raymond A.; Stucker, Donald E. 1985. Bighorn sheep and fire: seven case histories. In: Lotan, James E.; Brown, James K., comps. Fire's effects on wildlife habitat—symposium proceedings; 1984 March 21; Missoula, MT. Gen. Tech. Rep. INT-186. Ogden, UT: U.S. Department of Agriculture, Forest Service, Intermountain Research Station: 36-43.

Peek, J. M.; Riggs, R. A.; Lauer, J. L. 1979. Evaluation of fall burning on bighorn sheep winter range. Journal of Range Management. 32:430-432.

Perala, D. A. 1995. Quaking aspen productivity recovers after repeated prescribed fire. Res. Pap. NC-324. St. Paul, MN: U.S. Department of Agriculture, Forest Service, North Central Forest Experiment Station. 11 p.

Petersen, Kenneth L.; Best, Louis B. 1987. Effects of prescribed burning on nongame birds in a sagebrush community. Wildlife Society Bulletin. 15(3): 317-329.

Pfister, A. R. 1980. Post-fire avian ecology in Yellowstone National Park. Pullman, WA: Washington State University. Thesis. 35 p.

Pickering, Debbie L. 1997. The influence of fire on west coast grasslands and concerns about its use as a management tool: a case study of the Oregon silverspot butterfly Speyeria zerene Hippolyta (Lepidoptera, Nymphalidae). In: Greenlee, Jason M., ed. Proceedings, 1st conference on fire effects on rare and endangered species and habitats; 1995 November 13-16; Coeur d'Alene, ID. Fairfield, WA: International Association of Wildland Fire: 37-46.

Powell, Roger A.; Zielinski, William J. 1994. Fisher. In: Ruggiero, Leonard F.; Aubry, Keith B.; Buskirk, Steven W.; Lyon, L. Jack; Zielinski, William J., eds. American marten, fisher, lynx, and wolverine in the western United States: The scientific basis for conserving forest carnivores. Gen. Tech. Rep. RM-254. Fort Collins, CO: U.S. Department of Agriculture, Forest Service, Rocky Mountain Forest and Range Experiment Station: 38-73.

Prachar, Randy; Sage, R. W., Jr.; Deisch, M. S. 1988. Site occupancy, density, and spatial distribution of beaver colonies in burned and unburned areas in the Adirondacks. Transactions, Northeast Section of the Wildlife Society. 45: 74.

Probst, John R.; Weinrich, Jerry. 1993. Relating Kirtland's warbler population to changing landscape composition and structure. Landscape Ecology. 8(4): 257-271.

Provencher, L.; Galley, K. E. M.; Herring, B. J.; Sheehan, J.; Gobris, N. M.; Gordon, D. R.; Tanner, G. W.; Hardesty, J. L.; Rodgers, H. L.; McAdoo, J. P.; Northrup, M. N.; McAdoo, S. J.; Brennan, L. A. 1998. Post-treatment analysis of restoration effects on soils, plants, arthropods, and birds in sandhill systems at Eglin Air Force Base, Florida. Annual report to Natural Resources Division, Eglin Air Force Base, Niceville, FL. Gainesville, FL: Public Lands Program, The Nature Conservancy. 247 p.

Pulliam, H. R. 1988. Sources, sinks, and population regulation. American Naturalist. 132: 652-669.

Purcell, Alice; Schnoes, Roger; Starkey, Edward. 1984. The effects of prescribed burning on mule deer in Lava Beds National Monument. Corvallis, OR: Oregon State University, School of Forestry, Cooperative Park Studies Unit: 111-119.

Pyle, William H.; Crawford, John A. 1996. Availability of foods of sage grouse chicks following prescribed fire in sagebrush-bitterbrush. Journal of Range Management. 49(4): 320-324.

Pylypec, Bohdan. 1991. Impacts of fire on bird populations in a fescue prairie. Canadian Field-Naturalist. 105(3): 346-349.

Pyne, Stephen J. 1982. Fire in America, a cultural history of wildland and rural fire. Seattle, WA: University of Washington Press. 654 p.

Quinn, Ronald D. 1979. Effects of fire on small mammals in the chaparral. Cal-Neva Wildlife Transactions. 1979: 125-133.

Quinn, Ronald D. 1990. Habitat preferences and distribution of mammals in California chaparral. Res. Pap. PSW-202. Berkeley, CA: U.S. Department of Agriculture, Forest Service, Pacific Southwest Research Station. 11 p.

Raphael, M. G.; Morrison, M. L.; Yoder-Williams, M. P. 1987. Breeding bird populations during twenty-five years of postfire succession in the Sierra Nevada. Condor. 89: 614-626.

Ream, Catherine H., comp. 1981. The effects of fire and other disturbances on small mammals and their predators: an annotated bibliography. Gen. Tech. Rep. INT-106. Ogden, UT: U.S. Department of Agriculture, Forest Service, Intermountain Forest and Range Experiment Station. 55 p.

Reed, C. C. 1997. Responses of prairie insects and other arthropods to prescription burns. Natural Areas Journal. 17: 380-385.

Reich, P. B.; Abrams, M. D.; Ellsworth, D. S.; Kruger, E. L.; Tabone, T. J. 1990. Fire affects ecophysiology and community dynamics of central Wisconsin oak forest regeneration. Ecology. 71: 2179-2190.

Reichman, O. J. 1987. Konza Prairie: a tallgrass natural history. Lawrence, KS: University Press of Kansas. 226 p.

Reynolds, Richard T.; Graham, Russell T.; Reiser, M. Hildegard; Bassett, Richard L.; Kennedy, Patricia L.; Boyce, Douglas A., Jr.; Goodwin, Greg; Smith, Randall; Fisher, E. Leon. 1992. Management recommendations for the northern goshawk in the southwestern United States. Gen. Tech. Rep. RM-217. Fort Collins, CO: U.S. Department of Agriculture, Forest Service, Rocky Mountain Forest and Range Experiment Station. 90 p.

Rice, Lucile A. 1932. The effect of fire on the prairie animal communities. Ecology. 13(4): 392-401.

Riebold, R. J. 1971. The early history of wildfires and prescribed burning. In: Prescribed burning symposium: Proceedings. New Orleans, LA: U.S. Department of Agriculture, Forest Service, Southeastern Forest Experiment Station: 11-20.

Rieman, Bruce; Lee, Danny; Chandler, Gwynne; Myers, Deborah. 1997. Does wildfire threaten extinction for salmonids? Responses of redband trout and bull trout following recent large fires on the Boise National Forest. In: Greenlee, Jason M., ed. Proceedings, 1st conference on fire effects on rare and endangered species and habitats; 1995 November 13-16; Coeur d'Alene, ID. Fairfield, WA: International Association of Wildland Fire: 47-57.

70

USDA Forest Service Gen. Tech. Rep. RMRS-GTR-42-vol. 1. 2000

Riggs, Robert A.; Peek, James M. 1980. Mountain sheep habitat-use patterns related to post-fire succession. Journal of Wildlife Management. 44(4): 933-938.

Robbins, Louise E.; Myers, Ronald L. 1992. Seasonal effects of prescribed burning in Florida: a review. Misc. Pub. No. 8. Tallahassee, FL: Tall Timbers Research, Inc. 96 p.

Rogers, Garry F.; Steele, Jeff. 1980. Sonoran desert fire ecology. In: Stokes, Marvin A.; Dieterich, John H., tech. coords. Proceedings of the fire history workshop; 1980 October 20-24; Tucson, AZ. Gen Tech. Rep. RM-81. Fort Collins, CO: U.S. Department Agriculture, Forest Service, Rocky Mountain Forest and Range Experiment Station: 15-19.

Romme, W. H.; Despain, D. G. 1989. The long history of fire in the greater Yellowstone ecosystem. Western Wildlands. 15(2): 10-17.

Romme, William H. 1980. Fire frequency in subalpine forests of Yellowstone national Park. In: Stokes, Marvin A.; Dieterich, John H., tech. coords. Proceedings of the fire history workshop; 1980 October 20-24; Tucson, AZ. Gen Tech. Rep. RM-81. Fort Collins CO: U.S. Department Agriculture, Forest Service, Rocky Mountain Forest and Range Experiment Station: 27-30.

Rothermel, Richard C.; Hartford, Roberta A.; Chase, Carolyn H. 1994. Fire growth maps for the 1988 Greater Yellowstone Area fires. Gen. Tech. Rep. INT-304. Ogden, UT: U.S. Department of Agriculture, Forest Service, Intermountain Research Station. 64 p.

Rouse, Cary. 1986. Fire effects in northeastern forests: oak. Gen. Tech. Rep. NC-105. St. Paul, MN: U.S. Department of Agriculture, Forest Service, North Central Forest Experiment Station. 7 p.

Rowe, J. S. 1983. Concepts of fire effects on plant individuals and species. In: Wein, Ross W.; MacLean, David A., eds. The role of fire in northern circumpolar ecosystems. New York, NY: John Wiley and Sons: 135-153.

Rowland, M. M.; Alldredge, A. W.; Ellis, J. E.; Weber, B. J.; White, G. C. 1983. Comparative winter diets of elk in New Mexico. Journal of Wildlife Management. 47(4) :924-932.

Rundel, Philip W.; Parsons, David J. 1980. Nutrient changes in two chaparral shrubs along a fire-induced age gradient. American Journal of Botany. 67(1): 51-58.

Russell, Kevin R. 1999. [personal communication]. October 29. Dallas, OR: Willamette Industries, Inc.

Russell, Kevin R.; Van Lear, David H.; Guynn, David C., Jr. 1999. Prescribed fire effects on herpetofauna: review and management implications. Wildlife Society Bulletin. 27(2): 374-384.

Ryan, Kevin C.; Noste, Nonan V. 1985. Evaluating prescribed fires. In: Lotan, James E.; Kilgore, Bruce M.; Fischer, William C.; Mutch, Robert W., tech. coords. Proceedings—symposium and workshop on wilderness fire; 1983 November 15-18; Missoula, MT. Gen. Tech. Rep. INT-182. Ogden, UT: U.S. Department of Agriculture, Forest Service, Intermountain Forest and Range Experiment Station: 230-238.

Saab, Victoria A.; Dudley, Jonathan G. 1998. Responses of cavity-nesting birds to stand-replacement fire and salvage logging in ponderosa pine/Douglas-fir forests of southwestern Idaho. Res. Pap. RMRS-RP-11. Fort Collins, CO: U.S. Department of Agriculture, Forest Service, Rocky Mountain Research Station. 17 p.

Safford, L. O.; Bjorkbom, John C.; Zasada, John C. 1990. *Betula papyrifera* Marsh. paper birch. In: Burns, Russell M.; Honkala, Barbara H., tech. coords. Silvics of North America. Vol. 2. Hardwoods. Agric. Handb. 654. Washington, DC: U.S. Department of Agriculture, Forest Service: 158-171.

Sampson, Arthur W.; Jespersen, Beryl S. 1963. California range brushlands and browse plants. Berkeley, CA: University of California, Division of Agricultural Sciences, California Agricultural Experiment Station, Extension Service. 162 p.

Sando, R. W. 1978. Natural fire regimes and fire management—foundations for direction. Western Wildlands. 4(4): 35-44.

Schaefer, James A.; Pruitt, William O., Jr. 1991. Fire and woodland caribou in southeastern Manitoba. Wildlife Monographs. 116: 1-39.

Schardien, Bette J.; Jackson, Jerome A. 1978. Extensive ground foraging by pileated woodpeckers in recently burned pine forests. The Mississippi Kite. 8(1): 7-9.

Schiff, A. L. 1962. Fire and waters: scientific heresy in the Forest Service. Cambridge, MA: Harvard University Press. 225 p.

Schmid, J. M.; Thomas, L.; Rogers, T. J. 1981. Prescribed burning to increase mortality of Pandora moth pupae. Res. Note RM-405. Fort Collins, CO: U.S. Department of Agriculture, Forest Service, Rocky Mountain Forest and Range Experiment Station. 3 p.

Schmoldt, Daniel L.; Peterson, David L.; Keane, Robert E.; Lenihan, James M.; McKenzie, Donald; Weise, David R.; Sandberg, David V. 1999. Assessing the effects of fire disturbance on ecosystems: a scientific agenda for research and management. Gen. Tech. Rep. PNW-GTR-455. Portland, OR: U.S. Department of Agriculture, Forest Service, Pacific Northwest Research Station. 104 p.

Schroeder, M. H.; Sturges, D. L. 1975. The effect on the Brewer's sparrow of spraying big sagebrush. Journal of Range Management. 28: 294-297.

Schultz, Cheryl B.; Crone, Elizabeth E. 1998. Burning prairie to restore butterfly habitat: a modeling approach to management tradeoffs for the Fender's Blue. Restoration Ecology. 6(3): 244-252.

Schwartz, C. C.; Franzmann, A. W. 1989. Bears, wolves, moose and forest succession: Some management considerations on the Kenai peninsula. Alces. 25: 1-10.

Schwilk, Dylan W.; Keeley, Jon E. 1998. Rodent populations after a large wildfire in California chaparral and coastal sage scrub. Southwestern Naturalist. 43(4): 480-483.

Scott, Norman J., Jr. 1996. Evolution and management of the North American grassland herpetofauna. In: Finch, Deborah M., ed. Ecosystem disturbance and wildlife conservation in western grasslands, a symposium proceedings; 22-26 September 1994; Albuquerque, NM. Gen. Tech. Rep. RM-GTR-285. Fort Collins, CO: U.S. Department of Agriculture, Forest Service, Rocky Mountain Forest and Range Experiment Station: 40-53.

Seastedt, T. R.; Hayes, D. C.; Petersen, N. J. 1986. Effects of vegetation, burning and mowing on soil macroarthropods of tallgrass prairie. In: Clambey, Gary K.; Pemble, Richard H., eds. The prairie: past, present and future: Proceedings, 9th North American prairie conference; 1984 July 29-August 1; Moorhead, MN. Fargo, ND: Tri-College University Center for Environmental Studies: 99-102.

Seip, D. R.; Bunnell, F. L. 1985. Nutrition of Stone's sheep on burned and unburned ranges. Journal of Wildlife Management. 49(2): 397-405.

Severson, Kieth E.; Medina, Alvin L. 1983. Deer and elk habitat management in the Southwest. Journal of Range Management Monograph No. 2. 64 p.

Severson, Kieth E.; Rinne, John N. 1990. Increasing habitat diversity in southwestern forests and woodlands via prescribed fire. In: Krammes, J. S., tech. coord. Effects of fire management of southwestern natural resources: Proceedings; 1988 Nov. 15-17; Tucson, AZ. Gen. Tech. Rep. RM-191. Fort Collins, CO: U.S. Department of Agriculture, Forest Service, Rocky Mountain Forest and Range Experiment Station: 94-104.

Sharps, Jon C.; Uresk, Daniel W. 1990. Ecological review of black-tailed prairie dogs and associated species in western South Dakota. Great Basin Naturalist. 50(4): 339-345.

Shaw, James H.; Carter, Tracy S. 1990. Bison movements in relation to fire and seasonality. Wildlife Society Bulletin. 18(4): 426-430.

Sieg, Carolyn Hull; Severson, Kieth E. 1996. Managing habitats for white-tailed deer in the Black Hills and Bear Lodge Mountains of South Dakota and Wyoming. Gen. Tech. Rep. RM-GTR-274. Fort Collins, CO: U.S. Department of Agriculture, Forest Service, Rocky Mountain Forest and Range Experiment Station. 24 p.

Siemann, Evan; Haarstad, John; Tilman, David. 1997. Short-term and long-term effects of burning on oak savanna arthropods. American Midland Naturalist. 137 (2): 349-361.

Simons, Lee H. 1991. Rodent dynamics in relation to fire in the Sonoran Desert. Journal of Mammalogy. 72(3): 518-524.

Simovich, Marie A. 1979. Post fire reptile succession. Cal-Neva Wildlife Transactions. 1979: 104-113.

Singer, Francis J.; Schreier, William; Oppenheim, Jill; Garton, Edward O. 1989. Drought, fires, and large mammals. Bioscience. 39: 716-722.

Singer, Francis J.; Schullery, Paul. 1989. Yellowstone wildlife: populations in process. Western Wildlands. 15(2): 18-22.

Smallwood, John A.; Woodrey, Mark; Smallwood, Nathan J.; Kettler, Mary Anne. 1982. Foraging by cattle egrets and American kestrels at a fire's edge. Journal of Field Ornithology. 53(2): 171-172.

USDA Forest Service Gen. Tech. Rep. RMRS-GTR-42-vol. 1. 2000

71

Smith, Helen Y. 1999. Assessing longevity of ponderosa pine (*Pinus ponderosa*) snags in relation to age, diameter, wood density and pitch content. Missoula, MT: The University of Montana. Thesis. 46 p.

Smith, Jane Kapler; Fischer, William C. 1997. Fire ecology of the forest habitat types of northern Idaho. Gen. Tech. Rep. INT-GTR-363. Ogden, UT: U.S. Department of Agriculture, Forest Service, Intermountain Research Station. 142 p.

Spofford, Walter R. 1971. The golden eagle—rediscovered. Conservationist. 26(August-September): 6-8.

Stager, D. Waive.; Klebenow, Donald A. 1987. Mule deer response to wildfire in Great Basin pinyon-juniper woodland. In: Everett, Richard L., comp. Proceedings: pinyon-juniper conference; 1986 January 13-16; Reno, NV. Gen. Tech. Rep. INT-215. Ogden, UT: U.S. Department of Agriculture, Forest Service, Intermountain Research Station: 572-579.

Stanton, F. 1975. Fire impacts on wildlife and habitat. An abstracted bibliography of pertinent studies. U.S. Department of the Interior, Bureau of Land Management. Denver, CO: Denver Service Center. 48 p.

Stanton, P. A. 1986. Comparison of avian community dynamics of burned and unburned coastal sage scrub. Condor. 88: 285-289.

Stark, N.; Steele, R. 1977. Nutrient content of forest shrubs following burning. American Journal of Botany. 64(10): 1218-1224.

Stebbins, Robert C. 1985. A field guide to western reptiles and amphibians. Boston, MA: Houghton Mifflin Company. 336 p.

Stensaas, Mark. 1989. Forest fire birding. Loon. 61(1): 43-44.

Stoddard, H. L. 1931. The bobwhite quail: its habits, preservation and increase. New York, NY: Charles Scribner's and Sons. 559 p.

Stoddard, H. L. 1935. Use of controlled fire in southeastern upland game management. Journal of Forestry. 33: 346-351.

Stoddard, H. L. 1936. Relations of burning to timber and wildlife. Transactions, North American Wildlife Conference. 1: 399-403.

Stransky, John J.; Harlow, Richard F. 1981. Effects of fire on deer habitat in the Southeast. In: Wood, Gene W., ed. Prescribed fire and wildlife in southern forests: Proceedings of a symposium; 1981 Apr 6-8; Myrtle Beach, SC. Georgetown, SC: The Belle W. Baruch Forest Science Institute of Clemson University: 135-142.

Stubbendieck, James; Hatch, Stephan L.; Butterfield, Charles H. 1992. North American range plants. 4th ed. Lincoln, NE: University of Nebraska Press. 493 p.

Sutton, R. F.; Tinus, R. W. 1983. Root and root system terminology. Forest Science Monograph No. 24. 137 p.

Svedarsky, W. D.; Wolfe, T. J.; Kohring, M. A.; Hanson, L. B. 1986. Fire management of prairies in the prairie-forest transition of Minnesota. In: Koonce, Andrea L., ed. Prescribed burning in the Midwest: state-of-the-art: Proceedings of a symposium; 1986 March 3-6; Stevens Point, WI. Stevens Point, WI: University of Wisconsin-Stevens Point, College of Natural Resources, Fire Science Center: 103-107.

Svejcar, T. J. 1990. Response of *Andropogon gerardii* to fire in the tallgrass prairie. In: Collins, Scott L.; Wallace, Linda L., eds. Fire in North American tallgrass prairies. Norman, OK: University of Oklahoma Press: 18-27.

Sveum, Colin M.; Edge, W. Daniel; Crawford, John A. 1998. Nesting habitat selection by sage grouse in south-central Washington. Journal of Range Management. 51(3): 265-269.

Swain, A. M. 1973. A history of fire and vegetation in northeastern Minnesota as recorded in lake sediments. Quarternary Research. 3: 383-396.

Szeicz, J. M.; MacDonald, G. M. 1990. Postglacial vegetation history of oak savanna in southern Ontario. Canadian Journal of Botany. 69: 1507-1519.

Taber, Richard D.; Dasmann, Raymond F. 1958. The black-tailed deer of the chaparral. Game Bulletin No. 8. Sacramento, CA: State of California, Department of Fish and Game, Game Management Branch. 166 p.

Tappeiner, John; Zasada, John; Ryan, Peter; Newton, Michael. 1988. Salmonberry clonal and population structure in Oregon forests: the basis for a persistent cover. Unpublished paper on file at: College of Forestry, Oregon State University, Corvallis, OR: U.S. Department of Agriculture, Forest Service, Pacific Northwest Research Station. 28 p.

Taylor, Alan H; Halpern, Charles B. 1991. The structure and dynamics of *Abies magnifica* forests in the southern Cascade Range, USA. Journal of Vegetation Science. 2: 189-200.

Taylor, D. L. 1969. Biotic succession of lodgepole pine forests of fire origin in Yellowstone National Park. Laramie, WY: University of Wyoming. Dissertation. 320 p.

Taylor, Dale L. 1979. Forest fires and the tree-hole nesting cycle in Grand Teton and Yellowstone National Parks. In: Linn, R. M., ed. Proceedings of the 1st conference on scientific research in the National Parks; 1976 November 9-12; New Orleans, LA. Washington, DC: U.S. Department of Agriculture; National Park Service: 509-511.

Taylor, D. L.; Barmore, W. J., Jr. 1980. Post-fire succession of avifauna in coniferous forests of Yellowstone and Grand Teton National Parks, Wyoming. In: Workshop proceedings of the management of western forests and grasslands for nongame birds; 1980 February 11-14; Salt Lake City, UT. Gen. Tech. Rep. INT-86. Ogden, UT: U.S. Department of Agriculture, Forest Service, Intermountain Forest and Range Experiment Station: 130-145.

Telfer, E. S. 1993. Wildfire and the historical habitats of the boreal forest avifauna. In: Kuhnke, D. H., ed. Birds in the boreal forest, proceedings of a workshop; 1992 March 10-12; Prince Albert, SK. Catalogue No. Fo18-22/1992E. Edmonton, AB: Forestry Canada, Northwest Region, Northern Forestry Centre: 27-37.

Tewes, Michael E. 1984. Opportunistic feeding by white-tailed hawks at prescribed burns. Wilson Bulletin. 96(1): 135-136.

Thackston, Reginald E.; Hale, Philip E.; Johnson, A. Sydney; Harris, Michael J. 1982. Chemical composition of mountain-laurel *Kalmia* leaves from burned and unburned sites. Journal of Wildlife Management. 46(2): 492-496.

Thill, Ronald E.; Martin, Alton, Jr.; Morris, Hershel F., Jr.; McCune, E. Donice. 1987. Grazing and burning impacts on deer diets on Louisiana pine-bluestem range. Journal of Wildlife Management. 51(4): 873-880.

Thomas, D. C.; Barry, S. J.; Alaie, G. 1995. Fire-caribou-winter range relationships in northern Canada. Rangifer. 16(2): 57-67.

Thomas, Jack Ward; Forsman, Eric D.; Lint, Joseph B.; Meslow, E. Charles; Noon, Barry R.; Verner, Jared. 1990. A conservation strategy for the northern spotted owl. Report of the Interagency Scientific Committee to address the conservation of the northern spotted owl. Washington, DC: U.S. Government Printing Office. 427 p.

Thomas, P. A. 1991. Response of succulents to fire: a review. International Journal of Wildland Fire. 1(1): 11-22.

Thompson, Margaret W.; Shaw, Michael G.; Umber, Rex W.; Skeen, John E.; Thackston, Reggie E. 1991. Effects of herbicides and burning on overstory defoliation and deer forage production. Wildlife Society Bulletin. 19(2): 163-170.

Tomback, Diana F. 1986. Post-fire regeneration of krummholz whitebark pine: a consequence of nutcracker seed caching. Madrono. 33(2): 100-110.

Tomback, Diana F.; Carsey, Katherine S.; Powell, Mary L. 1996. Post-fire patterns of whitebark pine (*Pinus albicaulis*) germination and survivorship in the Greater Yellowstone area. In: Greenlee, Jason M., ed. The ecological implications of fire in Greater Yellowstone. Proceedings, 2nd biennial Conference on the Greater Yellowstone Ecosystem; 1993 September 19-21; Yellowstone National Park, WY. Fairfield, WA: International Association of Wildland Fire: 21.

Turner, M. G. 1990. Spatial and temporal analysis of landscape patterns. Landscape Ecology. 4(1): 21-30.

Turner, Monica G.; Hargrove, William W.; Gardner, Robert H.; Romme, William H. 1994. Effects of fire on landscape heterogeneity in Yellowstone National Park, Wyoming. Journal of Vegetation Science. 5: 731-742.

U.S. Department of Agriculture, Natural Resources Conservation Service. 1999. The PLANTS database. (http://plants.usda.gov/plants). Baton Rouge, LA: National Plant Data Center, 70874-4490 USA.

U.S. Department of the Interior. 1996. Effects of military training and fire in the Snake River Birds of Prey National Conservation Area. BLM/IDARING Research Project Final Report. Boise, ID: U.S. Geological Survey, Biological Resources Division, Snake River Field Station. 130 p.

U.S. Department of the Interior, Fish and Wildlife Service. 1985. Red-cockaded woodpecker recovery plan. Atlanta, GA: U.S. Fish and Wildlife Service. 88 p.

72

USDA Forest Service Gen. Tech. Rep. RMRS-GTR-42-vol. 1. 2000

U.S. Department of the Interior, Fish and Wildlife Service. 1995. Recovery plan for the Mexican spotted owl: Vol. I. Albuquerque, NM: U.S. Department of the Interior, Fish and Wildlife Service, Southwestern Region. 370 p.

Vacanti, P. Lynne; Geluso, Kenneth N. 1985. Recolonization of a burned prairie by meadow voles (*Microtus pennsylvanicus*). Prairie Naturalist. 17(1): 15-22.

Vales, David J.; Peek, James M. 1996. Responses of elk to the 1988 Yellowstone fires and drought. In: Greenlee, Jason M., ed. The ecological implications of fire in Greater Yellowstone. Proceedings, 2nd biennial conference on the Greater Yellowstone Ecosystem; 1993 September 19-21; Yellowstone National Park, WY. Fairfield, WA: International Association of Wildland Fire: 159-167.

Van Lear, David H. 1991. Fire and oak regeneration in the Southern Appalachians. In: Nodvin, Stephen C.; Waldrop, Thomas A., eds. Fire and the environment: ecological and cultural perspectives; 1990 March 20-24; Knoxville, TN. Gen. Tech. Rep. SE-69. Asheville, NC: U.S. Department of Agriculture, Forest Service, Southeastern Forest Experiment Station: 15-21.

Van Lear, David H.; Watt, Janet M. 1993. The role of fire in oak regeneration. In: Loftis, David L.; McGee, Charles E., eds. Oak Regeneration: serious problems, practical recommendations, symposium proceedings; 1992 Sep 8-10; Knoxville, TN. Gen. Tech. Rep. SE-84. Asheville, NC: U.S. Department of Agriculture, Forest Service, Southeastern Forest Experiment Station: 66-78.

Van Wagner, C. E. 1969. A simple fire-growth model. Forestry Chronicle. April: 103-104.

Van Wagner, C. E. 1978. Age-class distribution and the forest fire cycle. Canadian Journal of Forest Research. 8: 220-227.

Ver Steeg, Jeffrey M.; Harty, Francis M.; Harty, Lorree. 1983. Prescribed fire kills meadow voles (Illinois). Restoration and Management Notes. 1(4): 21.

Viereck, L. A. 1983. The effects of fire in black spruce ecosystems of Alaska and northern Canada. In: Wein, Ross W.; MacLean, David A., eds. The role of fire in northern circumpolar ecosystems. New York, NY: John Wiley and Sons: 201-220.

Viereck, L. A.; Dyrness, C. T. 1979. Ecological effects of the Wickersham Dome fire near Fairbanks, Alaska. Gen. Tech. Rep. PNW-90. Portland, OR: U.S. Department of Agriculture, Forest Service, Pacific Northwest Forest and Range Experiment Station. 71 p.

Vinton, Mary Ann; Harnett, David C.; Finck, Elmer J.; Briggs, John M. 1993. Interactive effects of fire, bison (*Bison bison*) grazing and plant community composition in tallgrass prairie. American Midland Naturalist. 129: 10-18.

Vogl, Richard J. 1967. Controlled burning for wildlife in Wisconsin. In: Proceedings, 6th annual Tall Timbers fire ecology conference; 1967 March 6-7; Tallahassee, FL. Tallahassee, FL: Tall Timbers Research Station: 47-96.

Vogl, Richard J. 1970. Fire and the northern Wisconsin pine barrens. In: Proceedings, 10th annual Tall Timbers fire ecology conference; 1970 August 20-21; Fredericton, NB. Tallahassee, FL: Tall Timbers Reserach Station: 175-209.

Waldrop, Thomas A.; Lloyd, F. Thomas. 1991. Forty years of prescribed burning on the Santee fire plots: effects on overstory and midstory vegetation. In: Nodvin, Stephen C.; Waldrop, Thomas A., eds. Fire and the environment: ecological and cultural perspectives; 1990 March 20-24; Knoxville, TN. Gen. Tech. Rep. SE-69. Asheville, NC: U.S. Department of Agriculture, Forest Service, Southeastern Forest Experiment Station: 45-59.

Waldrop, Thomas A.; Van Lear, David H.; Lloyd, F. Thomas; Harms, William R. 1987. Long-term studies of prescribed burning in loblolly pine forests of the southeastern coastal plain. Gen. Tech. Rep. SE-45. Asheville, NC: U.S. Department of Agriculture, Forest Service, Southeastern Forest Experiment Station. 23 p.

Ward, Peter. 1968. Fire in relation to waterfowl habitat of the delta marshes. In: Proceedings, 8th annual Tall Timbers fire ecology conference; 1968 March 14-15; Tallahassee, FL. Tallahassee, FL: Tall Timbers Research Station: 255-267.

Weatherspoon, C. Phillip; Husari, Susan J.; van Wagtendonk, Jan W. 1992. Fire and fuels management in relation to owl habitat in forests of the Sierra Nevada and southern California. In: Verner, Jared; McKelvey, Kevin S.; Noon, Barry R.; Gutierrez,

R. J.; Gould, Gordon I., Jr.; Beck, Thomas W., tech. coords. The California spotted owl: a technical assessment of its current status. Gen. Tech. Rep. PSW-GTR-133. Albany, CA: U.S. Department of Agriculture, Forest Service, Pacific Southwest Research Station: 247-260.

Weaver, H. 1943. Fire as an ecological and silvicultural factor in the ponderosa pine region of the Pacific slope. Journal of Forestry. 41: 7-15.

Weaver, H. 1974. Effects of fire on temperate forests: western United States. In: Kozlowski, T. T.; Ahlgren, C. E., eds. Fire and ecosystems. New York, NY: Academic Press: 279-319.

Wein, R. W. 1993. Historical biogeography of fire: circumpolar taiga. In: Crutzen, P. J.; Goldammer, J. G., eds. Fire in the environment. New York, NY: John Wiley and Sons: 267-276.

Wein, R. W.; Moore, J. M. 1977. Fire history and rotations in the New Brunswick Acadian Forest. Canadian Journal of Forest Research. 7: 285-294.

Welch, Bruce L.; Wagstaff, Fred J.; Williams, Richard L. 1990. Sage grouse status and recovery plan for Strawberry Valley, UT. Res. Pap. INT-430. Ogden, UT: U.S. Department of Agriculture, Forest Service, Intermountain Research Station. 10 p.

West, N. E.; Van Pelt, N. S. 1987. Successional patterns in pinyon-juniper woodlands. In: Everett, R. L., comp. Proceedings: pinyon-juniper conference; 1986 January 13-16; Reno, NV. Gen. Tech. Rep. INT-215. Ogden, UT: U.S. Department of Agriculture, Forest Service, Intermountain Research Station: 43-52.

West, Stephen D. 1982. Dynamics of colonization and abundance in central Alaskan populations of the northern red-backed vole, Clethrionomys rutilus. Journal of Mammalogy. 63(1): 128-143.

Whelan, Robert J. 1995. The ecology of fire. New York, NY: Cambridge University Press. 346 p.

Whisenant, Steven G. 1990. Changing fire frequencies on Idaho's Snake River plains: ecological and management implications. In: McArthur, E. Durant; Romney, Evan M.; Smith, Stanley D.; Tueller, Paul T., comps. Proceedings of a symposium on cheatgrass invasion, shrub die-off, and other aspects of shrub biology and management; 1989 April 5-7; Las Vegas, NV. Gen. Tech. Rep. INT-276. Ogden, UT: U.S. Department of Agriculture, Forest Service, Intermountain Research Station: 4-10.

White, David L.; Waldrop, Thomas A.; Jones, Steven M. 1991. Forty years of prescribed burning on the Santee fire plots: effects on understory vegetation. In: Nodvin, Stephen C.; Waldrop, Thomas A., eds. Fire and the environment: ecological and cultural perspectives; 1990 March 20-24; Knoxville, TN. Gen. Tech. Rep. SE-69. Asheville, NC: U.S. Department of Agriculture, Forest Service, Southeastern Forest Experiment Station: 51-59.

Wicklow-Howard, Marcia. 1989. The occurrence of vesicular-arbuscular mycorrhizae in burned areas of the Snake River Birds of Prey Area, Idaho. Mycotaxon. 34(1): 253-257.

Wiens, J. A.; Rotenberry, J. T. 1981. Habitat associations and community structure of birds in shrubsteppe environments. Ecological Monographs. 51: 21-41.

Williams, J.; Rich, R.; Cook, W.; Stephens, S. 1998. An evaluation of U.S. Forest Service wildfire acres burned trends. In: Viegas, D. X., ed. III International Conference on Forest Fire Research, 14th Conference on Fire and Forest meteorology; 16-20 November 1998; Luso, Coimbra, Portugal. Coimbra, Portugal: Associacao para o Desenvolvimento da Aerodinamica Industrial: 183-188.

Williams, M. 1989. Americans and their forests—a historical geography. New York, NY: Cambridge University Press. 599 p.

Wink, Robert L.; Wright, Henry A. 1973. Effects of fire on an Ashe juniper community. Journal of Range Management. 26(5): 326-329.

Winter, B. M.; Best, L. B. 1985. Effects of prescribed burning on placement of sage sparrow nests. Condor. 87: 294-295.

Wirtz, William O., II. 1977. Vertebrate post-fire succession. In: Mooney, Harold A.; Conrad, C. Eugene, tech. coords. Proceedings of the symposium on environmental consequences of fire and fuel management in Mediterranean ecosystems; 1977 Aug 1-5; Palo Alto, CA. Gen. Tech. Rep. WO-3. Washington, DC: U.S. Department of Agriculture, Forest Service: 46-57.

Wirtz, William O., II. 1979. Effects of fire on birds in chaparral. Cal-Neva Wildlife Transactions. 1979: 114-124.

Wittie, Roger D.; McDaniel, Kirk C. 1990. Effects of tebuthiuron and fire on pinyon-juniper woodlands in southcentral New Mexico. In:

Krammes, J. S., tech. coord. Effects of fire management of southwestern natural resources; 1988 Nov. 15-17; Tucson, AZ. Gen. Tech. Rep. RM-191. Fort Collins, CO: U.S. Department of Agriculture, Forest Service, Rocky Mountain Forest and Range Experiment Station: 174-179.

Witz, Brian W.; Wilson, Dawn S. 1991. Distribution of *Gopherus polyphemus* and its vertebrate symbionts in three burrow categories. American Midland Naturalist. 126(1): 152-158.

Wolff, Jerry O. 1978. Burning and browsing effects on willow growth in interior Alaska. Journal of Wildlife Management. 42(1): 135-140.

Woolfenden, Glen E. 1973. Nesting and survival in a population of Florida scrub jays. The Living Bird. 12: 25-49.

Wright, H. A. 1986. Effect of fire on arid and semi-arid ecosystems—North American continent. In: Joss, P. J.; Lynch, D. W.; Williams, D. B., eds. Rangelands under siege; Proceedings, International Rangeland Congress; 1984; Adelaide, Australia. New York, NY: Cambridge University Press: 575-576.

Wright, Henry A. 1973. Fire as a tool to manage tobosa grasslands. In: Proceedings, 12th annual Tall Timbers fire ecology conference; 1972 June 8-9; Lubbock, TX. Tallahassee, FL: Tall Timbers Research Station: 153-167.

Wright, Henry A. 1980. The role and use of fire in the semidesert grass-shrub type. Gen. Tech. Rep. INT-85. Ogden, UT: U.S. Department of Agriculture, Forest Service, Intermountain Forest and Range Experiment Station. 24 p.

Wright, Henry A.; Bailey, Arthur W. 1982. Fire ecology, United States and southern Canada. New York, NY: John Wiley and Sons. 501 p.

Wright, Vita. 1996. Multi-scale analysis of flammulated owl habitat use: owl distribution, habitat management, and conservation. Missoula, MT: University of Montana. Thesis. 90 p.

Yensen, Eric; Quinney, Dana L.; Johnson, Kathrine; Timmerman, Kristina; Steenhof, Karen. 1992. Fire, vegetation changes, and population fluctuations of Townsend's ground squirrels. American Midland Naturalist. 128: 299-312.

Yoakum, J. D.; O'Gara, B. W.; Howard, V. W., Jr. 1996. Pronghorn on western rangelands. In: Rangeland wildlife. Denver, CO: Society for Range Management: 211-226.

Yoakum, Jim. 1980. Habitat management guides for the American pronghorn antelope. Tech. Note 347. Denver, CO: U.S. Department of the Interior, Bureau of Land Management, Denver Service Center. 77 p.

Young, James A.; Evans, Raymond A. 1973. Downy brome—intruder in the plant succession of big sagebrush communities in the Great Basin. Journal of range Management. 26(6): 410-415.

Zasada, John; Tappeiner, John; Maxwell, Bruce. 1989. Manual treatment of salmonberry or which bud's for you? Cope Report, Coastal Oregon Productivity Enhancement Program. 2(2): 7-9.

Zimmerman, J. L. 1992. Density-independent factors affecting the avian diversity of the tallgrass prairie community. Wilson Bulletin. 104: 85-94.

74

USDA Forest Service Gen. Tech. Rep. RMRS-GTR-42-vol. 1. 2000

Appendices

Appendix A: Common and Scientific Names of Animal Species _____

Taxonomy for birds is from Ehrlich and others (1988); for reptiles and amphibians, Conant and Collins (1991) and Stebbins (1985); for mammals, Jones and others (1992); for insects, Borror and White (1970).

Common name	Scientific name
Agile kangaroo rat	*Dipodomys agilis*
American Kestrel	*Falco sparverius*
American marten	*Martes americana*
American Robin	*Turdus migratorius*
American Wigeon	*Anas americana*
American badger	*Taxidea taxus*
American beaver	*Castor canadensis*
Arachnids (includes spiders and ticks)	*Arachnidae*
Bachman's Sparrow	*Aimophila aestivalis*
Bald Eagle	*Haliaeetus leucocephalus*
Barn Swallow	*Hirundo rustica*
Beetles	*Coleoptera*
Bewick's Wren	*Thryomanes bewickii*
Bighorn sheep	*Ovis canadensis*
Bison	*Bison bison*
Black bear	*Ursus americanus*
Black Vulture	*Coragyps atratus*
Black-backed Woodpecker	*Picoides arcticus*
Black-chinned Sparrow	*Spizella atrogularis*
Black-footed ferret	*Mustela nigripes*
Black-tailed jackrabbit	*Lepus californicus*
Blue Jay	*Cyanocitta cristata*
Blue-headed Vireo	*Vireo solitarius*
Blue-winged Teal	*Anas discors*
Box turtle	*Terrapene carolina*
Brewer's Sparrow	*Spizella breweri*
Brown Creeper	*Certhia americana*
Brown-headed Nuthatch	*Sitta pusilla*
Brush mouse	*Peromyscus boylii*
Brush rabbit	*Sylvilagus bachmani*
Bugs	*Hemiptera*
Burrowing Owl	*Athene cunicularia*
California Gnatcatcher	*Polioptila californica*
California Quail	*Callipepla californica*
California mouse	*Peromyscus californicus*
California pocket mouse	*Chaetodipus californicus*
Canyon Towhee	*Pipilo fuscus*
Caribou	*Rangifer tarandus*
Carolina Wren	*Thryothorus ludovicianus*
Cattle Egret	*Bubulcus ibis*
Central Florida crowned snake	*Tantilla relicta*
Chigger	*Acarina*
Chipping Sparrow	*Spizella passerina*
Cicadas, hoppers, whiteflies, aphids, scale insects	*Homoptera*

USDA Forest Service Gen. Tech. Rep. RMRS-GTR-42-vol. 1. 2000

75

Appendix A

Common name	Scientific name
Clark's Nutcracker	*Nucifraga columbiana*
Clay-colored Sparrow	*Spizella pallida*
Common Grackle	*Quiscalus quiscula*
Common gray fox	*Urocyon cinereoargenteus*
Common Ground-dove	*Columbina passerina*
Common Raven	*Corvus corax*
Common Yellowthroat	*Geothlypis trichas*
Cooper's Hawk	*Accipiter cooperii*
Cotton rat	*Sigmodon* spp.
Coyote	*Canis latrans*
Crested Caracara	*Caracara plancus*
Crossbills	*Loxia* spp.
Dall's sheep	*Ovis dalli*
Dark-eyed Junco	*Junco hyemalis*
Darkling beetles	*Tenebrionoidea*
Deer mouse	*Peromyscus maniculatus*
Desert woodrat	*Neotoma lepida*
Desert tortoise	*Gopherus agassizii*
Diamondback rattlesnake	*Crotalus admanteus*
Downy Woodpecker	*Picoides pubescens*
Dusky-footed woodrat	*Neotoma fuscipes*
Eastern glass lizard	*Ophisaurus ventralis*
Eastern Kingbird	*Tyrannus tyrannus*
Eastern Wood-pewee	*Contopus virens*
Eastern Bluebird	*Sialia sialis*
Eastern Towhee	*Pipilo erythrophthalmus*
Elk	*Cervus elaphus*
European Starling	*Sturnus vulgaris*
Feather Mite	*Acarina*
Fender's blue butterfly	*Icaricia icarioides fenderi*
Ferruginous Hawk	*Buteo regalis*
Field Sparrow	*Spizella pusilla*
Flammulated Owl	*Otus flammeolus*
Flatwoods salamander	*Ambystoma singulatum*
Florida Scrub-Jay	*Aphelocoma coerulescens coerulescens*
Fowler's toad	*Bufo woodhousii*
Fox Sparrow	*Passerella iliaca*
Golden eagle	*Aquila chrysaetos*
Golden-crowned Kinglet	*Regulus satrapa*
Gopher tortoise	*Gopherus polyphemus*
Gopher snake	*Pituophis melanoleucus*
Grasshopper Sparrow	*Ammodramus savannarum*
Grasshoppers, katydids, crickets, mantids, walkingsticks, and cockroaches	*Orthoptera*
Gray wolf	*Canis lupus*
Gray Catbird	*Dumetella carolinensis*
Great Horned Owl	*Bubo virginianus*
Great Crested Flycatcher	*Myiarchus crinitus*
Great Blue Heron	*Ardea herodias*
Greater Prairie-Chicken	*Tympanuchus cupido pinnatus*
Grizzly bear	*Ursus arctos*
Ground squirrel	*Spermophilus* spp.

Appendix A

Common name	Scientific name
Hairy Woodpecker	*Picoides villosus*
Hammond's Flycatcher	*Empidonax hammondii*
Heath Hen	*Tympanuchus cupido cupido*
Heerman kangaroo rat	*Dipodomys heermanni*
Hermit Thrush	*Catharus guttatus*
Hooded Warbler	*Wilsonia citrina*
Indigo Bunting	*Passerina cyanea*
June beetles	*Melolonthinae*
Kangaroo rat species	*Dipodomys* spp.
Karner blue butterfly	*Lycaeides melissa samuelis*
Kirtland's Warbler	*Dendroica kirtlandii*
Lark Sparrow	*Chondestes grammacus*
Lark Bunting	*Calamospiza melanocorys*
Lazuli Bunting	*Passerina amoena*
Lewis' Woodpecker	*Melanerpes lewis*
Loggerhead Shrike	*Lanius ludovicianus*
Lynx	*Lynx lynx*
Meadow vole	*Microtus pennsylvanicus*
Mole skink	*Eumeces egregius*
Moose	*Alces alces*
Mountain lion	*Felis concolor*
Mountain goat	*Oreamnos americanus*
Mountain Bluebird	*Sialia currucoides*
Mourning Dove	*Zenaida macroura*
Mule deer	*Odocoileus hemionus*
Northern Flicker	*Colaptes auratus*
Northern flying squirrel	*Glaucomys sabrinus*
Northern Goshawk	*Accipiter gentilis*
Northern Harrier	*Circus cyaneus*
Northern red-backed vole	*Clethrionomys rutilus*
Northern Shoveler	*Anas clypeata*
Northern Bobwhite	*Colinus virginianus*
Northern Cardinal	*Cardinalis cardinalis*
Nuthatches	*Sitta* spp.
Ord's kangaroo rat	*Dipodomys ordii*
Ovenbird	*Seiurus aurocapillus*
Peregrine Falcon	*Falco peregrinus*
Pileated Woodpecker	*Dryocopus pileatus*
Pine Warbler	*Dendroica pinus*
Plains pocket gopher	*Geomys bursarius*
Pocket mouse species	*Perognathus* spp.
Pocket gopher species	*Thomomys, Geomys* spp.
Prairie dog	*Cynomys ludovicianus*
Prairie Falcon	*Falco mexicanus*
Pronghorn	*Antilocapra americana*
Rabbit	*Sylvilagus* spp.
Raccoon	*Procyon lotor*
Red squirrel	*Tamiasciurus hudsonicus*
Red fox	*Vulpes vulpes*
Red-bellied Woodpecker	*Melanerpes carolinus*
Red-breasted Nuthatch	*Sitta canadensis*
Red-cockaded Woodpecker	*Picoides borealis*
Red-eyed Vireo	*Vireo olivaceus*

USDA Forest Service Gen. Tech. Rep. RMRS-GTR-42-vol. 1. 2000

77

Appendix A

Common name	Scientific name
Red-naped Sapsucker	*Sphyrapicus nuchalis*
Red-shouldered Hawk	*Buteo lineatus*
Red-spotted newt	*Notophthalmus viridescns*
Red-tailed Hawk	*Buteo jamaicensis*
Red-winged Blackbird	*Agelaius phoeniceus*
Ruffed Grouse	*Bonasa umbellus*
Sage Grouse	*Centrocercus urophasianus*
Sage Sparrow	*Amphispiza belli*
Sage Thrasher	*Oreoscoptes montanus*
Sand skink	*Neoseps reynoldsi*
Sapsucker species	*Sphyrapicus* spp.
Savannah Sparrow	*Passerculus sandwichensis*
Scrub-Jay	*Aphelocoma* spp.
Sharp-shinned Hawk	*Accipiter striatus*
Sharp-tailed Grouse	*Tympanuchus phasianellus*
Short-eared Owl	*Asio flammeus*
Shrew species	*Sorex* and *Blarina* spp.
Six-lined racerunner	*Cnemidophorus sexlineatus*
Snowshoe hare	*Lepus americanus*
Southern harvester ant	*Pogonomyrmex badius*
Spotted Owl	*Strix occidentalis*
Spruce budworm	*Choristoneura fumiferana*
Spruce Grouse	*Falcipennis canadensis*
Summer Tanager	*Piranga rubra*
Swainson's Thrush	*Catharus ustulatus*
Three-toed Woodpecker	*Picoides tridactylus*
Townsend's chipmunk	*Tamias townsendii*
Townsend's ground squirrel	*Spermophilus townsendii*
Tree Swallow	*Tachycineta bicolor*
Turkey Vulture	*Cathartes aura*
Upland Sandpiper	*Bartramia longicauda*
Vaux's Swift	*Chaetura vauxi*
Vagrant shrew	*Sorex vagrans*
Vole species	*Microtus, Clethrionomys,* and *Phenacomys* spp.
Western Bluebird	*Sialia mexicana*
Western fence lizard	*Sceloporus occidentalis*
Western harvest mouse	*Reithrodontomys megalotis*
Western Meadowlark	*Sturnella neglecta*
Western rattlesnake	*Crotalus viridis*
Western Screech-Owl	*Otus kennicottii*
Western Tanager	*Piranga ludoviciana*
White-eyed Vireo	*Vireo griseus*
White-headed Woodpecker	*Picoides albolarvatus*
White-tailed deer	*Odocoileus virginianus*
White-tailed Hawk	*Buteo albicaudatus*
White-throated woodrat	*Neotoma albigula*
Wild Turkey	*Meleagris gallopavo*
Wood Thrush	*Hylocichla mustelina*
Woodland salamanders	*Desmognathus aeneus, Desmognathus ochrophaeus, Eurycea wilderae, Plethodon jordani*
Woodrat species	*Neotoma* spp.
Wrentit	*Chamaea fasciata*
Yellow-rumped Warbler	*Dendroica coronata*

Appendix B: Common and Scientific Names of Plant Species_____

Names are from U.S. Department of Agriculture, Natural Resources Conservation Service (1999).

Common name	Scientific name
alder	*Alnus* spp.
alligator juniper	*Juniperus deppeana*
American beech	*Fagus grandifolia*
antelope bitterbrush	*Purshia tridentata*
Ashe's juniper	*Juniperus ashei*
balsam fir	*Abies balsamea*
barrel cactus	*Ferocactus* spp.
big bluestem	*Andropogon gerardii*
big sagebrush	*Artemisia tridentata*
bigleaf maple	*Acer macrophyllum*
bigtooth aspen	*Populus grandidentata*
bitter cherry	*Prunus emarginata*
black grama	*Bouteloua eriopoda*
black spruce	*Picea mariana*
blackgum	*Nyssa sylvatica*
blueberry	*Vaccinium* spp.
bluebunch wheatgrass	*Pseudoroegneria spicata*
bluegrass	*Poa* spp.
bluestem	*Andropogon* spp.
buckbrush	*Ceanothus cuneatus*
bur oak	*Quercus macrocarpa*
California red fir	*Abies magnifica*
cattail	*Typha* spp.
chamise	*Adenostoma fasciculatum*
Chapman oak	*Quercus chapmanii*
cheatgrass	*Bromus tectorum*
chinkapin oak	*Quercus muehlenbergii*
cholla	*Opuntia fulgida*
deerbrush	*Ceanothus integerrimus*
Douglas-fir	*Pseudotsuga menziesii*
eastern redcedar	*Juniperus virginiana*
eastern white pine	*Pinus strobus*
eastern hemlock	*Tsuga canadensis*
fir	*Abies* spp.
Gambel oak	*Quercus gambelii*
Geyer's sedge	*Carex geyeri*
giant sequoia	*Sequoiadendron giganteum*
grand fir	*Abies grandis*
gray birch	*Betula populifolia*
greenleaf manzanita	*Arctostaphylos patula*
hickory	*Carya* spp.
huckleberry (western species)	*Vaccinium* spp.
Idaho fescue	*Festuca idahoensis*
incense cedar	*Calocedrus decurrens*
Indian ricegrass	*Achnatherum hymenoides*
Indiangrass	*Sorghastrum nutans*
jack pine	*Pinus banksiana*
Jeffrey pine	*Pinus jeffreyi*
juniper	*Juniperus* spp.
loblolly pine	*Pinus taeda*
lodgepole pine	*Pinus contorta*

USDA Forest Service Gen. Tech. Rep. RMRS-GTR-42-vol. 1. 2000

79

Appendix B

Common name	Scientific name
longleaf pine	*Pinus palustris*
manzanita	*Arctostaphylos* spp.
mesquite	*Prosopis* spp.
mullein species	*Verbascum* spp.
myrtle oak	*Quercus myrtifolia*
northern red oak	*Quercus rubra*
oak	*Quercus* spp.
oneseed juniper	*Juniperus monosperma*
paper birch	*Betula papyrifera*
pinegrass	*Calamagrostis rubescens*
pinyon pines	*Pinus cembroides, P. edulis, P. monophylla*
pitch pine	*Pinus rigida*
plains reedgrass	*Calamagrostis montanensis*
pond pine	*Pinus serotina*
ponderosa pine	*Pinus ponderosa*
prairie dropseed	*Sporobolus heterolepis*
pricklypear	*Opuntia* spp.
quaking aspen	*Populus tremuloides*
red alder	*Alnus rubra*
red huckleberry	*Vaccinium parvifolium*
red maple	*Acer rubrum*
red pine	*Pinus resinosa*
red spruce	*Picea rubens*
redwood	*Sequoia sempervirens*
rough dropseed	*Sporobolus clandestinus*
rough fescue	*Festuca altaica* (subspecies *F. hallii, F. campestris*)
runner oak	*Quercus margarettiae*
sagebrush	*Artemisia* spp.
salal	*Gaultheria shallon*
salmonberry	*Rubus spectabilis*
sand live oak	*Quercus geminata*
sand pine	*Pinus clausa*
saw palmetto	*Serenoa repens*
sawgrass	*Cladium* spp.
Scouler's willow	*Salix scouleriana*
sedge species	*Carex* spp.
shadbush	*Amelanchier arborea*
shortleaf pine	*Pinus echinata*
silverberry	*Eleagnus commutata*
Sitka spruce	*Picea sitchensis*
slash pine	*Pinus elliottii*
southern magnolia	*Magnolia grandiflora*
spruce species	*Picea* spp.
spurge species	*Euphorbia* spp.
subalpine fir	*Abies lasiocarpa*
sugar maple	*Acer saccharum*
sugar pine	*Pinus lambertiana*
sweetgum	*Liquidambar styraciflua*
thickspike wheatgrass	*Elymus macrourus*
thimbleberry	*Rubus parviflorus*
tobosagrass	*Pleuraphis mutica*
tuliptree	*Liriodendron tulipifera*
turkey oak	*Quercus laevis*

Appendix B

Common name	Scientific name
Utah juniper	*Juniperus osteosperma*
velvet mesquite	*Prosopis velutina*
vine maple	*Acer circinatum*
wedgeleaf ceanothus	*Ceanothus cuneatus*
western hemlock	*Tsuga heterophylla*
western juniper	*Juniperus occidentalis*
western larch	*Larix occidentalis*
western snowberry	*Symphoricarpos occidentalis*
western wheatgrass	*Pascopyrum smithii*
white fir	*Abies concolor*
white spruce	*Picea glauca*
whitebark pine	*Pinus albicaulis*
wild lupine	*Lupinus perennis*
willow species	*Salix* spp.
winterfat	*Krascheninnikovia lanata*
yellow birch	*Betula alleghaniensis*

USDA Forest Service Gen. Tech. Rep. RMRS-GTR-42-vol. 1. 2000

81

Appendix C: Glossary

The definitions here were derived from the following: fuels and fire behavior from Agee (1993), Brown and others (1982), Helms (1998), National Park Service and others (1998), Ryan and Noste (1985); fire occurrence from Agee (1993), Johnson (1992), and Romme (1980); plant reproduction from Allaby (1992), Sutton and Tinus (1983); other terms from Lincoln and others (1998).

abundance: The total number of individuals of a species in an area or community.

climax: A biotic community that is in equilibrium with existing environmental conditions and represents the terminal stage of an ecological succession.

cohort: A group of individuals of the same age, recruited into a population at the same time; age class.

connectivity: Accessibility of suitable habitat from population centers. All patches of suitable habitat that can be reached and occupied are considered connected.

crown fire: Fire that burns in the crowns of trees and shrubs, usually ignited by a surface fire. Crown fires are common in coniferous forests and chaparral shrublands.

density: The number of individuals within a given area.

dominance (dominant): The extent to which a given species predominates in a community because of its size, abundance, or coverage.

duff: Partially decomposed organic matter lying beneath the litter layer and above the mineral soil. It includes the fermentation and humus layers of the forest floor (02 soil horizon).

duration of fire: The length of time that combustion occurs at a given point. Relates closely to downward heating and fire effects below the fuel surface as well as heating of tree boles above the surface.

fire cycle: Used in this volume as equivalent to **fire rotation.**

fire exclusion: The policy of suppressing all wildland fires in an area.

fire frequency: A general term referring to the recurrence of fire in a given area over time. Sometimes stated as number of fires per unit time in a designated area. Also used to refer to the probability of an element burning per unit time.

fire intensity: Used in this volume as equivalent to **fireline intensity.**

fire regime: General pattern of fire frequency, season, size, and prominent, immediate effects in a vegetation type or ecosystem.

fire return interval: Number of years between fires at a given location.

fire rotation: The length of time necessary for an area equal in size to the study area to burn.

fire severity: A qualitative measure of the immediate effects of fire on the ecosystem. Relates to the extent of mortality and survival of plant and animal life both above and below ground and to loss of organic matter.

fireline intensity: The rate of energy release per unit length of the fire front expressed as BTU per foot of fireline per second or as kilowatts per meter of fireline. This expression is commonly used to describe the power of wildland fires.

flame length: The length of flames in the propagating fire front measured along the slant of the flame from the midpoint of its base to its tip. Mathematically related to fireline intensity and the height of scorch in the tree crown.

fuel: Living and dead vegetation that can be ignited. For descriptions of kinds of fuels and fuel classification, see "Effects of Fire on Flora" in the Rainbow Series.

fuel continuity: A qualitative description of the distribution of fuel both horizontally and vertically. Continuous fuels readily support fire spread. The larger the fuel discontinuity, the greater the fire intensity required for fire spread.

fuel loading: Weight per unit area of fuel often expressed in tons per acre or tonnes per hectare. Dead woody fuel loadings are commonly described for small material in diameter classes of 0 to 1/4-, 1/4 to 1-, and 1 to 3-inches and for large material in one class greater than 3 inches.

ground fire: Fire that burns in the organic material below the litter layer, mostly by smoldering combustion. Fires in duff, peat, dead moss, lichens, and partly decomposed wood are typically ground fires.

herpetile: Amphibian or reptile.

ladder fuels: Shrubs and young trees that provide continuous fine material from the forest floor into the crowns of dominant trees.

litter: The top layer of the forest floor (01 soil horizon); includes freshly fallen leaves, needles, fine twigs, bark flakes, fruits, matted dead grass, and a variety of miscellaneous vegetative parts that are little altered by decomposition. Litter also accumulates beneath rangeland shrubs. Some surface feather moss and lichens are considered to be litter because their moisture response is similar to that of dead fine fuel.

mast: Fruits of all flowering plants used by wildlife, including fruits with fleshy exteriors (such as berries) and fruits with dry or hard exteriors (such as nuts and cones).

mean fire return interval: The arithmetic average of all fire intervals in a given area over a given time period.

mesic: Pertaining to conditions of moderate moisture or water supply.

mixed severity fire regime: Regime in which fires either cause selective mortality in dominant vegetation, depending on different species' susceptibility to fire, or vary between understory and stand replacement.

prescribed fire: Any fire ignited by management actions to meet specific objectives. Prior to ignition, a written, approved prescribed fire plan must exist, and National Environmental Protection Act requirements must be met.

presettlement fire regime: The time from about 1500 to the mid- to late-1800s, a period when Native American populations had already been heavily impacted by European presence but before extensive settlement by European Americans in most parts of North America, before extensive conversion of wildlands for agricultural and other purposes, and before fires were effectively suppressed in many areas.

rhizome: A creeping stem, not a root, growing beneath the surface; consists of a series of nodes with roots commonly produced from the nodes and producing buds in the leaf axils.

scatter-hoard: Seed cached in scattered shallow holes, a common behavior for kangaroo rats and pocket mice.

secondary cavity nester: Animal that lives in tree cavities but does not excavate them itself.

sere: A succession of plant communities leading to a particular association.

snag: A standing dead tree from which the leaves and some of the branches have fallen.

stand replacement fire regime: Regime in which fires kill or top-kill aboveground parts of the dominant vegetation, changing the aboveground structure substantially. Approximately 80 percent or more of the aboveground dominant vegetation is either consumed or dies as a results of fires. Applies to forests, shrublands, and grasslands.

succession: The gradual, somewhat predictable process of community change and replacement, leading toward a climax community; the process of continuous colonization and extinction of populations at a particular site.

surface fire: Fire that burns in litter and other live and dead fuels at or near the surface of the ground, mostly by flaming combustion.

top-kill: Kills aboveground tissues of plant without killing underground parts from which the plant can produce new stems and leaves.

total heat release: The heat released by combustion during burnout of all fuels in BTU per square foot or kilocalories per square meter.

underburn: Understory fire.

understory fire regime: Regime in which fires are generally not lethal to the dominant vegetation and do not substantially change the structure of the dominant vegetation. Approximately 80 percent or more of the aboveground dominant vegetation survives fires. Applies to forest and woodland vegetation types.

wildland fire: Any nonstructure fire, other than prescribed fire, that occurs in a wildland.

xeric: Having very little moisture; tolerating or adapted to dry conditions.

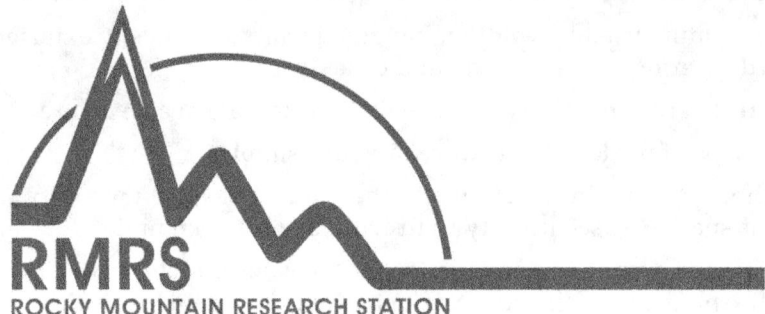

RMRS
ROCKY MOUNTAIN RESEARCH STATION

The Rocky Mountain Research Station develops scientific information and technology to improve management, protection, and use of the forests and rangelands. Research is designed to meet the needs of National Forest managers, Federal and State agencies, public and private organizations, academic institutions, industry, and individuals.

Studies accelerate solutions to problems involving ecosystems, range, forests, water, recreation, fire, resource inventory, land reclamation, community sustainability, forest engineering technology, multiple use economics, wildlife and fish habitat, and forest insects and diseases. Studies are conducted cooperatively, and applications may be found worldwide.

Research Locations

Flagstaff, Arizona	Reno, Nevada
Fort Collins, Colorado*	Albuquerque, New Mexico
Boise, Idaho	Rapid City, South Dakota
Moscow, Idaho	Logan, Utah
Bozeman, Montana	Ogden, Utah
Missoula, Montana	Provo, Utah
Lincoln, Nebraska	Laramie, Wyoming

*Station Headquarters, Natural Resources Research Center, 2150 Centre Avenue, Building A, Fort Collins, CO 80526